The History of Science and Technology in the North West

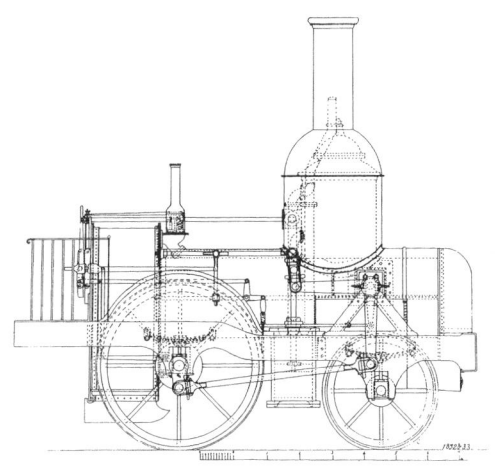

Manchester Region History Review
Volume 18
2007

ISSN 0952–4320

© Copyright Manchester Centre for Regional History, 2007

All rights are reserved and no part of the publication can be reprinted without the permission of the editors and authors.

Typeset by Carnegie Publishing, Lancaster
Printed and bound by Alden Press, Witney, Oxford

Contents

Contributor's Notes   vii

Editorial   ix
  John Pickstone

Articles

  John V. Pickstone
  Science and technology in Manchester: an introduction
  to the history   1

  Allan Chapman
  Under a Lancashire heaven: William Crabtree, Jeremiah
  Horrocks and their circle, and the origins of research
  astronomy in seventeenth-century England   19

  Richard Hills
  Richard Roberts (1789–1864), pioneer of production
  engineering in Manchester   41

  Graeme Gooday
  Cosmos, climate and culture: Manchester meteorology
  made universal   64

  Tim Cooper
  The early development of scientific research in industry:
  the case of Metropolitan-Vickers Ltd, 1901–1933   84

  Mary Jo Nye
  Manchester friends at odds: Michael Polanyi, P. M. S. Blackett
  and the scientist as political speaker   106

  Samuel J. M. M. Alberti
  Molluscs, mummies and moon rock: the Manchester Museum
  and Manchester science   130

Archives and Libraries

  Jan Hargreaves
  The Collections Centre at the Museum of Science
  and Industry, Manchester   155

Museums
   Francis Neary
   Manchester Science Gallery at the Museum of Science
   and Industry, Manchester   162

Societies
   Kevin Kilburn
   Manchester Astronomical Society   169

Long Reviews   171

The Short Reviews have been held over due to space limitations.

Editors
Melanie Tebbutt
Craig Horner
John Wilson

Editorial board

Steve Fielding	Chris Makepeace (Short Reviews)
Ruth Frow	Paula Moorhouse
Morris Garratt (Libraries)	John Pickstone
Clare Hartwell	Catharine Rew (Museums)
Karen Hunt	Mike Rose
Alan Kidd	Bill Williams
Neville Kirk	Terry Wyke
Brian Maidment	

Book Reviews editor
Craig Horner

Corresponding members
Robert Glen, University of New Haven, USA
Kazuhiko Kondo, University of Tokyo, Japan

Correspondence
The Editors
Manchester Region History Review
c/o Manchester Centre for Regional History
Manchester Metropolitan University
Geoffrey Manton Building
Rosamond Street West
Manchester M15 6LL
United Kingdom
http://www.mcrh.mmu.ac.uk

For full details of individual and institutional subscription rates for the United Kingdom and overseas, refer to
http://www.mcrh.mmu.ac.uk/pubs/mrhr.htm

Illustrations
We are grateful to the following for their help and permission in reproducing illustrations:
Mary Jo Nye
Ken Irving
Alan White
Robin Marshall
The University Librarian and Director, The John Rylands University Library, The University of Manchester
Every effort has been made to contact the copyright holders but if any have been inadvertently overlooked, the editors will be pleased to make the necessary arrangements at the first opportunity.

Notes for contributors
If you would like to contribute to this journal, please contact the editors before submitting copy. Authors should consult http://www.mcrh.mmu.ac.uk/pubs/guidelines.doc
Conventional articles should not exceed 8,000 words including footnotes, although they can be much shorter. We encourage a variety of contributions and are willing to discuss ideas and draft articles at an early stage. Intending contributors to the Libraries, Museums and Societies sections should consult the editors in the first instance. Book reviews should be sent to the Book Reviews editor. All submitted work should be typewritten.

Advertisements
For details of advertising rates please contact the editors.

Contributor's Notes

Sam Alberti is Lecturer at the University of Manchester Centre for Museology and Research Fellow at the Manchester Museum. Trained as a historian of science, he has published on the scientific culture of Victorian Britain, and on museums of anatomy and natural history. He is currently researching a history of the Manchester Museum. Sam is closely involved with the Museums and Galleries History Group, and Secretary of the University of Manchester Collection Curators' Forum.

Allan Chapman is a native of Manchester and a historian of science at Oxford University. His main areas of research include the history of astronomy and medicine, as well as the relationship between science and Christianity. He is the author of eight books and many articles, and is also involved in broadcasting.

Tim Cooper completed his PhD at the Centre for the History of Science, Technology and Medicine (CHSTM) in June 2003. His studies focus on the relations between Manchester University and local industry, 1914-1960, in the furtherance of scientific research. He currently works for the UN in Kosovo.

Graeme Gooday is a Senior Lecturer in History and Philosophy of Science at the University of Leeds. His specialist area of research is late-nineteenth century physics and electrical engineering, and in 2004 he published his first book *The morals of measurement: accuracy, irony and trust in late-Victorian electrical practice*.

Jan Hargreaves is the Senior Archivist at the Museum of Science and Industry in Manchester. Jan is responsible for the management of the archive collections at MSIM, and the operation of the museum's pioneering Collections Centre.

Richard Hills was a research assistant on the history of textile technology from 1965 to 1968. During this time he collected the first exhibits for the Manchester Museum of Science and Technology which opened in 1969 with him as Director. This led to the installation of a working self-acting mule and power loom, both based on Richard Roberts's inventions. An ill-health retirement in 1985 enabled him to help as a priest in the Church of England as well as write engineering history.

Kevin Kilburn, FRAS, is Secretary of the Manchester Astronomical Society.

Francis Neary is a Research Associate in the Centre for the History of Science, Technology and Medicine at the University of Manchester. His recent research has been concerned with the history of twentieth-century medical technologies and he has curated exhibitions on joint replacement at The Royal College of Surgeons of England, and Wrightington Hospital, near Wigan.

Mary Jo Nye is Horning Professor of the Humanities and Professor of History at Oregon State University in Corvallis, Oregon, USA. Her most recent book is *Blackett: physics, war, and politics in the twentieth century* (Harvard University Press, 2004), and she is currently working on a study of the scientific life and philosophy of science of Michael Polanyi.

John Pickstone comes from Burnley and has worked in Manchester since 1974. In 1986, at the University of Manchester, he founded the Wellcome Unit and the Centre for the History of Science, Technology and Medicine, where his present research is mostly on recent medicine, especially cancer services, medical technology, and the NHS around Manchester. He is the author of *Ways of knowing: a new history of science, technology and medicine* (Manchester University Press, 2000), and editor, with Roger Cooter, of the *Companion to medicine in the twentieth century* (Routledge World Reference, 2002).

This volume is dedicated to the memory of Donald Cardwell and Wilfred Farrar, astute and wry historians of science, technology and Manchester

Editorial

John Pickstone

My first and happy task as editor of this issue is to thank the contributors for their knowledge and skill, their ready co-operation, and their tolerance of the delays. There is, I think, a special sort of companionship which links historians of a particular place, and for a place as important as Manchester these links extend world-wide.

I am also most grateful to Craig Horner for his skilled editing and for the friendly spirit in which the work has been done. So also to Melanie Tebbutt for guidance and support, and more generally to the historians at Manchester Metropolitan University for their maintenance of this journal and for their researches on local and regional history since the days of Bill Williams and Manchester Studies in the 70s. We are fortunate indeed to have this focus for regional history.

Several of the following articles, and much of the literature I survey in the first essay, have derived from the Centre for the History of Science, Technology and Medicine (CHSTM) at the University of Manchester, and from its forerunner, the Department of History of Science and Technology at UMIST, established by Donald Cardwell in 1962. Much of the work of Cardwell, Arnold Pacey, Wilfred and Kathleen Farrer, and later of Joe Marsh and Raj Williamson was focussed on Manchester. Especially notable are the achievements of one of our present contributors, Richard Hills, who in the 60s and 70s established the Museum of Science and Industry in the Odd Fellows Hall on Grosvenor Street, and then led the re-establishment of the Museum at Castlefield. Its present success is a great tribute to Richard's work and to the support he received from the University, the City, and especially from Donald Cardwell and Lord Bowden at UMIST.

The work on local history of science, technology and medicine has continued since 1986 in our Centre (CHSTM) at the University of Manchester, now directed by my colleague Professor Michael Worboys. It is part of a much wider programme which focuses chiefly on the twentieth century and includes many different approaches and sites. Ongoing local work includes contributions to an Illustrated History of the University of Manchester which will appear in 2007; whilst Sam Alberti (see below) is completing a history of the Manchester Museum, and Duncan Wilson a history of the recent transformations in Manchester Life Sciences. In all this local work we try to be global in comparative reach and in historiographical inspiration. We are always ready to help students and scholars with a serious interest in Manchester STM, or in HSTM more generally.

It is also a pleasure to note that this present volume on the history of science and technology is a companion to a collection on the history of medicine in and around Manchester, a special number of the *Bulletin of the John Rylands Library of the University of Manchester*, vol. 87 (2007), edited with Stella Butler.

John Pickstone
Centre for the History of Science, Technology and Medicine

Caring for the Community.

Each year AstraZeneca will commit substantial funds and resources to a wide variety of community projects. The company has long been part of the community and has played its part in generating a thriving business environment.

In addition it has recognised the needs of less fortunate groups, helping with both funding and human resources.

Those who continue to benefit from this community spirit include schools, charities, the arts, sport, the elderly and the sick.

Alderley Park
50 years
in the community

AstraZeneca
life inspiring ideas

AP7867

Centre for the History of Science, Technology and Medicine

The University of Manchester's Centre for the History of Science, Technology and Medicine (CHSTM) is an international leader in research and teaching in the history of nineteenth- and twentieth-century science, technology and medicine. One of the most dynamic departments in this field in the UK, it has a large postgraduate community and an active and friendly research environment with a lively seminar programme. In addition to the superb John Rylands University Library, CHSTM has its own research library, and hosts the National Archive for the History of Computing. We have good connections with major museums of HSTM in Manchester and beyond. Many CHSTM staff have expertise in the history of science, technology and medicine in Manchester and the north-west, and we encourage local studies.

CHSTM offers taught Master's degrees in History of Science and Technology; History of Medicine; and Science Communication. MSc programmes last 1 year full-time (2 years part-time) and provide research training culminating in a dissertation based on original research under specialist supervision. More advanced research can be undertaken via the MPhil programme (1 year full-time, 2 years part-time) or the PhD (3 years full-time, 6 years part-time).

CHSTM also maintains a large and diverse postdoctoral research community and regularly hosts visiting scholars from the UK and overseas.

For further details of CHSTM, its staff and its teaching and research programmes, see www.manchester.ac.uk/chstm.

**We welcome enquiries from potential students and visitors.
Contact Dr Jeff Hughes, CHSTM, Simon Building, The University of Manchester, Oxford Road, Manchester M13 9PL. Tel: 0161 275 5857/5850.
Email: jeff.hughes@manchester.ac.uk.**

ARTICLES
Science and technology in Manchester: an introduction to the history

John V. Pickstone

Perhaps better than any other city, Manchester illustrates the historical relationship between the growth of science and the growth of industry – a relationship which is topical now, as it has been for two centuries. This introduction provides a sketch map and some references for those who would explore this local and regional history.[1]

In the sixteenth century, Manchester was a market town in a far province. Interest in mathematics, natural history or natural philosophy was then rather rare. The first 'name' in Manchester was John Dee, the Elizabethan mathematician, alchemist, geographer and necromancer who for some years around 1600 was Warden of the College attached to the parish church.[2] In the mid-seventeenth century, as the mathematical arts and the collecting of specimens became more common among the gentry, clergy and townsmen, Manchester was one site in a network of enthusiasts which extended north to Burnley and across the Pennines to Halifax. As we see below in Allan Chapman's article, the skills of a land surveyor such as William Crabtree could be extended from Salford to the heavens, and those grown used to accuracy on earth could also ask for accuracy above. Instruments, both practical or philosophical, were central to this work, and in the later seventeenth century the Towneley circle around Burnley were much concerned with barometers and the improvement of clocks.[3]

We know relatively little about how such concerns played out in southern Lancashire in the early eighteenth century, but more about a later network of enthusiasts around Kendal and the western Yorkshire Dales, one which extended to the West Indies, Edinburgh and London, as well as to the cities of the north west.[4] Many of this group were Quakers and the schools they ran were crucial; this was the background of John Dalton, before he came from the Lakes to Manchester. Many were doctors, like the Fothergills who succeeded in London, and John Haygarth who worked in Chester and who corresponded with James Currie in Liverpool and Thomas Percival in Manchester. (In the early nineteenth century, three academics of national importance came from this district: Adam Sedgwick, son of the parson in Dentdale and then Cambridge Professor of geology; William Whewell, the Lancaster carpenter's son who became master of Trinity College, Cambridge; and Richard Owen, also from Lancaster, who achieved fame for studies of fossils and became the first Director of the Museum of Natural History in South Kensington.[5])

The Manchester Enlightenment

By the 1770s, Manchester was booming as a centre of the textile trade and as a regional capital, and it is still possible around King Street and St Ann's Square to get the feel of Georgian Manchester. This was the 'better end' of town when polite society developed around the Assembly Rooms and the churches and the chapels. It was a society of gentlefolk, but especially of merchants and professional men.

The Manchester Infirmary in Piccadilly was the main focus for established and incoming doctors. Its leading surgeon, Charles White, was an authority on midwifery and a noted teacher. Its leading physician, Thomas Percival, was a key member of the Unitarian Congregation at Cross Street Chapel (now in a multi-storey office). With his minister there, and with his colleague Thomas Henry (the leading apothecary and manufacturing chemist), Percival established a scientific society which has continued to the present as the Manchester Literary and Philosophical Society.[6] He also helped establish the Manchester College which gave higher education to laymen as well as to future Unitarian ministers.[7] At the end of the century, as Manchester's urban problems grew severe and fever threatened, Percival and his colleagues addressed the problems of public health and fought for an expansion of the Infirmary to provide better facilities for infectious diseases. They had their successes, especially around 1790, but their projects were increasingly overtaken by the long years of war and repression which followed the French Revolution.[8]

It was Percival and his friends who first advanced schemes for higher education for young industrialists and professionals. They brought John Dalton to Manchester, as a teacher of chemistry and natural philosophy. It was here that the young Quaker asked himself why the atmosphere did not separate into the elements that it was now believed to contain; he reflected on the solubility of gases which his friend Thomas Henry was forcing into proprietary mineral waters; he thought about chemical combination, and about explaining chemistry to the young. It is to this college teacher, befriended by industrialists and by enthusiasts for Newtonian science and rational amusement, that we owe the atomic theory in chemistry. His book, *A new system of chemistry and philosophy*, was published in 1808. It was Dalton, as a scientific hero, who maintained the Literary and Philosophical Society through the difficult decades which opened the new century.[9]

Science and the classic industrial city

By the 1830s Manchester was attracting tourists who came to see the future. They visited the mills and the Infirmary to see textile production and its victims. They noted the cultural monuments which a newly assertive middle class had provided since the 1820s. Most of them were around Mosley Street and St Peter's Fields, the site of Peterloo, the political demonstration of 1819 which had become a massacre. The Natural History Society was for botanists and for lovers of country sports; the Royal Manchester Institution (whose building was designed by Charles Barry and is now the City Art Gallery) concentrated on art; the Athenaeum (now part of the Gallery extension) later catered for the younger gentlemen, many of whom found natural philosophy rather heavy.[10] The Mechanics Institute was established by engineers and businessmen to provide science for the working classes, the kind of education which Richard Roberts and other self-taught engineers had struggled to gain, or done without.[11] But, in fact some of the most devoted students proved to be clerks learning languages and accountancy. Most of the new societies had their own buildings, which often included museums and sometimes included a laboratory or workroom.[12] The Literary and Philosophical Society persisted for those of the middle classes sufficiently devoted to science as not to mind Dr Dalton talking yet again about meteorology. The keener members were often the engineers and chemists, who during the day were busy analysing chemicals and machines, improving production and sometimes looking to synthesise new combinations.[13]

In this foul yet exciting city, political economy and religion were the major topics of debate. Younger professionals and industrialists worried about the moral and physical condition of the working classes; their pamphlets argued for reform, for more information, for more middle class involvement – lest unrest and disease came to seriously damage the new industrial system. Hence the Manchester Statistical Society[14] and the tradition of sanitary reform; hence the missions and the savings banks; hence the Mechanics Institutes where workers might come to understand the laws of nature. Chemists and medical men, geologists and 'statisticians' collaborated in sanitary science.[15]

By the 1840s there was more scientific expertise to draw on. Some of the local manufacturers employed chemists who had trained in the new German university laboratories. Chemists were useful to analyse ingredients and products, so improving efficiency. Engineers in the larger local companies could use mathematical and experimental expertise: William Fairbarn asked Eaton Hodgkinson to help measure the strength of iron structures, and Hodgkinson went on to teach engineers in the new University of London.[16] J. F. Bateman, a major civil engineer, learned his science around the Lit and Phil,[17] as most importantly did James Prescott Joule, a pupil of Dalton and son of a local brewer. Amidst the mechanical and electrical enthusiasm of the early Victorian years, Joule's careful, conservative spirit sought security in measurement. He wrecked wild hopes of infinite power from electricity, but he established the mechanical equivalent of heat. Joule brought to physics the engineers' concepts of duty and efficiency. William Thomson, in Glasgow, took up his experiments and helped establish the principle of conservation of energy.[18]

The statues of Dalton and Joule now face each other in the porch of the gothic Town Hall – fine monuments to a largely amateur scientific culture in which intellect was directed not only to the regularities of nature but to the operations of men upon it.

Owens College and the Technical School

By the 1850s Manchester's central district was fully commercial. The Infirmary was remodelled, losing its sprawl of allied charities, gaining a portico and clock tower as a solid civic monument. On the streets around Piccadilly, the warehouses were becoming grander, now that the railways brought potential buyers to see the 'wares' collected and elaborately displayed. One of the merchants, John Owens, was persuaded by another to leave his fortune to provide a college for young men. This was not to be a sectarian affair like that newly established by the Congregationalists in Whalley Range; it was not to be tainted with the Unitarian heresy, like the Manchester New College, now back in its home city after 37 years in York.[19] Owens College was to be like Oxford and Cambridge in what it taught, but it would be nonsectarian and non-residential. It was established in 1851 in the former house of Richard Cobden, merchant, liberal MP, apostle of free trade, anti-militarist and phrenologist.[20]

It almost failed. Manchester fathers would go to occasional lectures, and they would turn out for a big fashionable event like the national Art Show held near Trafford Park in 1857. Some of them reckoned there might be commercial benefit in a short course on calico printing or some such; but they were not going to divert their sons into three years of university education. Owens College survived and grew because it came to incorporate a more vigorous form of symbiosis between capital and culture, one learned from Scotland and especially from Germany. The year of the College's opening, 1851, was also the year of the Great Exhibition in London; British industry was then supreme, but so was German science, and from Prince Albert downward there were educated men who feared that German success in science would lead to more powerful competition in industry. Their answer was to develop in Britain the system of advanced training and research which had been pioneered in German universities. In Manchester it was Henry Roscoe, Professor of chemistry from 1857, who developed that vision. He was well placed to do so: he was the son of a distinguished Liverpool family, a pupil of Professor Bunsen of Heidelberg, and he married into the Potter family, wealthy liberal merchants and major local politicians. He also recruited his cousin, the former chemist Stanley Jevons, to teach political economy.[21] Roscoe linked scientific research to the practical concerns of local industrialists, but always insisting that no one could be a useful chemist until he had mastered the principles of the subject; you could not start with applications.[22] When Owens College moved in 1873 to its present site on Oxford Road, Roscoe built a large chemical laboratory at the rear of the main building, developing one of the best chemistry departments in the world, with particular strengths in organic chemistry. Manchester then contained many industrial chemists, including several working on dyestuffs, but some returned to Germany and used university contacts to build a new industry. In both senses, they left behind the British companies for which they had worked.[23]

Henry Roscoe (1833–1915), Professor of chemistry at Owens College from 1857

It was Roscoe and his students who helped arrange the Charter of 1880 by which the Owens College became the first college of the Victoria University, later to be joined by colleges in Leeds and Liverpool.[24] By then the local medical school had been brought into the College; its staff included such notable 'scientific clinicians' as Julius Dreschfeld, born in Manchester but educated in Germany. Engineering classes were established under Osborne

Reynolds, and Horace Lamb's applied mathematics was to add further strengths.[25] The physics department had a good reputation for teaching – several of its pupils, including J. J. Thomson and Arthur Schuster, went on to the new Cavendish Laboratories at Cambridge. As Graeme Gooday shows in this volume, physics was then closely related to engineering, meteorology and chemistry: Balfour Stewart, Reynolds and later Schuster were all interested in 'outdoor physics'.[26]

In the 1880s the Natural History Society's collection, which had occupied their Museum near St Peter's Square, was rehoused as part of the new College. That did not mean that the Museum lost contact with the wider community; rather as Sam Alberti shows below, it was the link between the university and local naturalists, and also a shop window to attract attention and support.[27] From 1851, W. C. Williamson, a former curator of the Manchester Museum, and later a doctor, had taught geology as well as zoology and botany. A new curator, W. Boyd Dawkins, became Professor of geology in 1874, and in 1879 Arthur Milnes Marshall, a talented Cambridge embryologist, was given the Chair of zoology. Marshall was one of the followers of T. H. Huxley, both in popularising science and in creating a new science of biology focused on functions, tissues and cells rather than classifications. He died young and his successor in zoology was Sidney Hickson, an expert on corals, who supported women's education and built up biology as a school subject and as a university degree in its own right.

Williamson remained in the Chair of botany until 1891. Like much of the work at Owens, his research was linked to the local community and he was the major British pioneer in the study of coal fossils. He was succeeded in 1892 by Frederick Ernest Weiss, then aged 26. Born too in Britain of a German family like both Dreschfeld and Schuster, it was evident how much Manchester University owed to industrial England's mercantile connections with German learning. For four decades, Weiss and Hickson represented classical botany and zoology, together with W. H. Lang, who was appointed in 1909 to a new chair in cryptogamic botany. He too was an expert on plant fossils, which remained a Manchester specialism. The most famous of the fossil-botanists was Marie Stopes, appointed in 1904 and the first woman lecturer in the Faculty of Science, but better known later as an advocate of birth control.[28]

The Mechanics Institute had developed in parallel with Owens. A new building was opened in 1856 in Princess Street (the site of the first Trade Union Congress, and now housing the Labour Party archives). The occasion was marked with an international exhibition of Arts and Manufactures, but enrolments were disappointing and the development of elementary education under the municipal School Boards

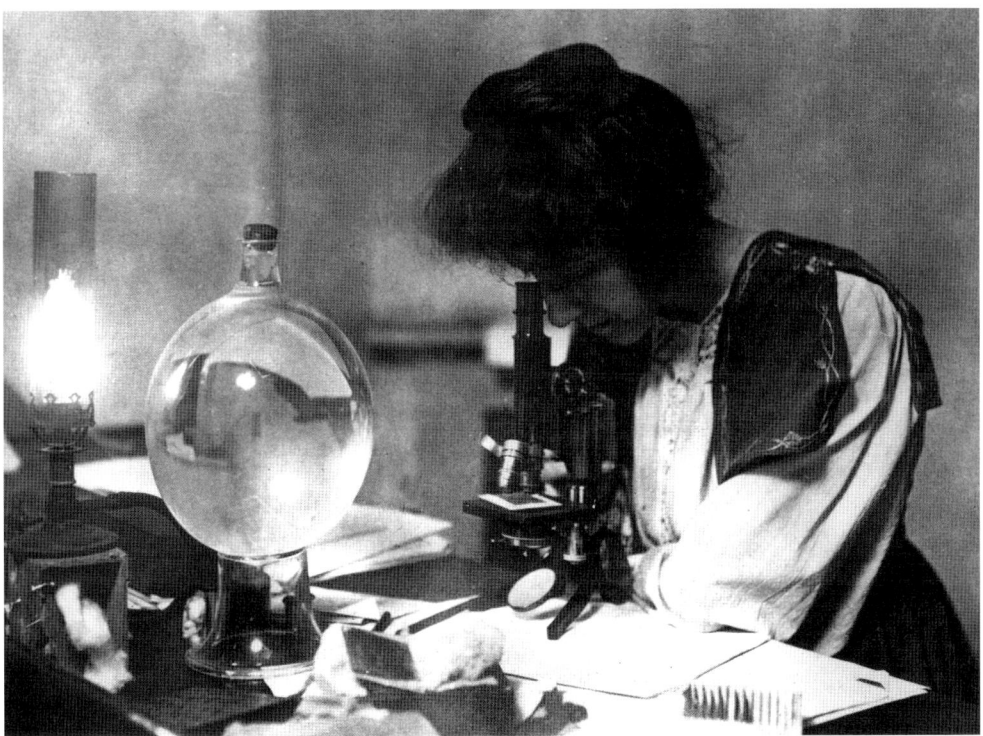

Marie Stopes (1880–1958), first female lecturer in the Faculty of Science, Victoria University

from the 1870s robbed the Institute of some of its previous functions. It was a night-school-educated boot-maker, J. H. Reynolds, who rescued it. Appointed its Secretary in 1879, he made it part of a growing national movement for technical education, linking it with the new City and Guilds examination, and soliciting industrial support. As British manufacturers worried more and more about international competition, and as the elementary schools gave a basic education to more of the working classes, so the Technical School grew. By 1889 a new building was planned as part of a new Whitworth Institute, which together with a School of Art and an Art Gallery, was to be funded from the legacy of the engineer Joseph Whitworth. But new legislation was passed, national funding became available to the 'great towns', and the Technical School was taken over by the city of Manchester in 1892. A fine new building of German inspiration was opened in 1901 on land provided from Whitworth monies (later the UMIST main building). The 'Tech' was intended to teach science for industrial application; Owens would teach the professional men, there being an agreement to that effect from 1896. But as money and equipment poured into the new Tech, the boundaries became blurred.[29]

The University around 1900

By about 1900, Owens College had achieved the ideals of Roscoe and his fellows. It was a 'Centre of Intelligence' for a major manufacturing region, and it had expanded its role in professional training. The medical school had been much enlarged and engineering courses were coming to be accepted as a partial substitute for apprenticeships. Owens was helping to produce teachers for the state elementary schools and for the technical colleges then being developed by the better municipalities. After a long struggle, women students had been fully recognised.

But colleges like Owens were coming to offer more; they were introducing scientific and technical novelties – the world of electricity, of X-rays and radioactivity, of bacteria and the causes of disease. In these new fields the colleges worked with local industry and local authorities, for whom they could provide analysis or testing facilities. It was in these fields especially that principles and practices were very close. The Whitworth Laboratories were opened in 1887 for engineering, and in 1894 the medical school was much enlarged. At about the same time the new Professor of pathology, Sheridan Delépine, opened a public health laboratory that served much of the region.[30] Physics got a splendid new laboratory in 1900, also including electro-technics and electro-chemistry, which gained their own building in 1912, three years after the old engineering laboratories had been replaced with much bigger facilities. In 1908, the Infirmary moved from the centre of town to become a neighbour of the University.[31] At this time, too, national legislation raised the standards for pharmaceutical education, and after the Great War the University was at last able to compete successfully with the private schools nearby.[32]

The capital for new buildings and new chairs had come mainly from industrialists. Local authority finance, which became important from about 1890, was directed chiefly to the Technical College as we have noted.[33] As it came to rival Owens in the higher reaches of engineering research and teaching, the reaction was mixed. Some university purists preferred to leave practical matters to the Technical College. But Schuster, now the Professor of physics, backed by colleagues in Arts, urged collaboration in the expansion of technical education. His arguments prevailed, and in 1903 certain senior staff in the Municipal College were recognised as constituting the Faculty of Technology of the University of Manchester, now newly independent from the colleges in Leeds and Liverpool.[34]

Similar tensions between practical training and university science were evident in medicine. The clinical professors who dominated the medical faculty were the elite local practitioners for whom teaching

Ernest Rutherford (1871–1937), Chair of physics at Victoria University from 1907

was a part-time activity; they and their students had little time for research, though the physician William Roberts had run a lab at his home. The professors of pre-clinical science, notably physiology, found little encouragement and little time for anything beyond undergraduate teaching and routine administration. Bacteriology was the major exception; that work was so marketable that pressure of routine testing tended to stifle original laboratory work.[35]

A neglect of research was common in British medical schools. Only in Liverpool, for instance, was medical science allied with local pharmaceutical companies, as happened more frequently in Germany and America. It is characteristic of Manchester around 1900 that its greatest achievements in the nascent fields of biochemistry and biotechnology came not from the medical school but from the chemistry department. It was there that Arthur Harden trained before he achieved fame at the Lister Institute in London for work on yeast;[36] it was there that Chaim Weizmann, future President of Israel, experimented with the acetone-butanol fermentation which proved vital to munitions production in the Great War.[37] It was chemists from this department, working with Manchester Corporation, who hit upon the activated sludge method of sewage treatment.[38]

In the physical sciences generally, researchers were in control. The chemistry department poured out research papers and also appealed to industry. Physics, under Ernest Rutherford, became the world centre for research on the structure of the atom; the list of his staff and students can still impress even the non-specialist. No one claimed that this (indoor and experimental) physics was useful, but it was enormously exciting.[39] And if you had visited the meteorological station on the moors above Glossop, you might have met Ludwig Wittgenstein, then a research student in engineering.[40]

Whether Wittgenstein knew Rutherford in Manchester we do not know, but Rutherford and the Danish physicist Neils Bohr were members of a reading group run by Samuel Alexander, the Australian-born philosopher who seems to have been the intellectual hub of the

University's increasingly cosmopolitan staff. Other associates included Grafton Elliot Smith, the Australian anatomist; Weizmann, the Russian chemist and Zionist; T. S. Tout, the (British) medievalist; Horace Lamb, the applied mathematician, who had served time in Australia; and Stopes, the young palaeobotanist.[41]

Before and after World War II

Edwardian Manchester was German in its culture – in its science as well as in philosophy and music; the Great War wrecked that attachment.[42] Soon afterwards came the long decline of the cotton trade, the former basis of Manchester's prosperity. The towns around were hard-hit, though Manchester itself was buffered somewhat by the diversity of its industries and service functions. Capital drained from the district and the epitaph of Lancashire cotton came to be written on gravestones far from the remnants of industry.

The inter-war decades were difficult for the University and for the Technical College. The Tech eventually planned an extension, but it was not completed until the 1950s. At Owens, the new buildings between the wars were largely for the arts departments, but the sciences continued to impress. After the Great War, the departments of botany and zoology lost the 'economic' biology posts they had struggled to develop; the state funding went south, like some key local industries.[43]

The immediate post-war years saw a national round of promotions: young men were getting their chances. In 1919 Elliot Smith went to University College London and was replaced as Professor of anatomy by the young John Stopford. William Stirling at last retired from physiology and the university appointed A. V. Hill, a young Cambridge graduate (and friend of Lawrence Bragg) whose work on muscle heat was to earn a Nobel prize. The medical faculty considered creating full-time chairs in medicine, surgery and obstetrics, but the proposal was deferred. Manchester tried but failed to get Edward Mellanby as a clinical researcher; he saw the Manchester Royal Infirmary as 'unreformed' and went instead to the small but research-minded faculty at Sheffield. In 1922, Henry Dean, the Professor of pathology since 1915, moved to Cambridge; and the following year Hill went to University College London. Their successors as medical scientists were the physiological chemist Henry Raper and the bacteriologists W. W. C. Topley and then H. B. Maitland. They were distinguished national figures but they did not draw international talent.

But organic chemistry thrived under Robert Robinson and then Alexander Todd, and physical chemistry under Michael Polanyi. The atomic physics of Rutherford gave way to the crystallography of

Lawrence Bragg, and then to the cosmic physics of P. M. S. Blackett. D. R. Hartree taught applied mathematics, theoretical physics, and pioneered computing techniques. In all these sciences, Manchester, with Cambridge or Oxford, helped support British communities which held their own with the damaged traditions of Germany and the rising institutions of the United States. As Mary Jo Nye shows in this volume, Polanyi and Blackett came to take very different views of the relationships of science with government. They became national leaders in opposed campaigns but remained the best of friends.[44]

It was perhaps in the practical, professional fields that Manchester's reputation rose most. At the Tech, Miles Walker developed electrical engineering in association with Metropolitan-Vickers; and Willis Jackson, in the Faculty of Science, continued this association around World War II.[45] As Tim Cooper shows in his article, Metro-Vicks in the 1930s was also closely associated with Cambridge, and it saw scientific research as a way of generating new products to keep it going through the depression; John Cockroft, who had studied electrical engineering at the Tech, was crucial to the Cambridge collaboration (and later to Britain's World War II atomic project).[46] In medicine, Stopford, though now Vice-Chancellor, was unable to convert the major clinical chairs to full-time appointments, but in the new specialities, where regional organisations mattered, Manchester medicine benefited from strong local-authority support. Harry Platt in orthopaedics, Geoffrey Jefferson in neurosurgery, and Ralston Paterson at the Christie cancer hospital achieved international reputations. In many of these fields, as in local industry, the models were often American.[47]

Manchester's tradition in chemistry, physics and engineering, plus its size and regional base, meant that it was well-placed to benefit from government projects after World War II. In many respects these set the agenda which persisted through the 1970s: radio-astronomy, from the war-time radar studies; computing from the work at Bletchley on code breaking; nuclear physics and engineering from the bomb projects; pharmaceutical manufacture, as Britain finally caught up with American and European patterns; and not least, the National Health Service, with state support for clinical research and for marginal medical specialities such as occupational health and rheumatology. In all these areas Manchester was able to achieve new reputations. Bernard Lovell's radio-telescope at Jodrell Bank became a national symbol for space exploration. The world's first stored-program computer, built here by F. C. Williams and Tom Kilburn, led to close collaboration with Ferranti and then ICL; the project also drew on the very strong department of mathematics where Max Newman had collected several of his Bletchley co-workers, including Alan Turing.[48] Nuclear engineering was a Manchester speciality, linked with the several nuclear installations developed in this region. The local dyestuffs industry developed, belatedly, into ICI Pharmaceuticals (now part of Astra-Zeneca) which had University links. At the Medical School, the establishment of full-time clinical chairs brought Robert Platt to Manchester to head a Department of Medicine, also graced by Douglas Black. These were all formidable initiatives, important aspects of national developments.

The 1950s and 1960s brought considerable expansion of staff and students in British universities. Manchester remained a strong provincial university, though no longer so exceptional. Expansion was particularly notable in the Faculty of Technology which in 1956, under Vivian Bowden, became a full university institution, UMIST. (The Polytechnic, later Manchester Metropolitan University, would take the non-degree work.) Manchester produced some big science and some big scientific statesmen, prominent on the new research councils and often moving on to London.[49]

From the 1980s to Project Unity

The post-war University (and UMIST) was dominated by the physical sciences. Caught between them and the strong clinical interests of the medical school, the bio-medical sciences were relatively weak. They were reorganised from 1983 at the instigation of Sir Mark Richmond, a new Vice-Chancellor who was himself a medical scientist, aided by a small group of young professors. The School of Biological Sciences which they created has been one of the recent success stories; without this upgrading Manchester would have struggled to maintain its national position at a time of formal research assessment and competition.[50]

By the end of the millennium, Manchester science and technology ranked with the other big provincial universities, and with some of the best of the newer universities. The region's industrial base was now much reduced: engineering was largely defence-related, the nuclear industry was shrinking and unpopular, and the computer firms that had merged as ICL were now owned in Japan, though Britain was good at software.[51] Of the post-war supports, only pharmaceuticals was really prosperous; and in a country which depended increasingly on services rather than manufacturing, physical sciences were decreasingly popular amongst university entrants. Undergraduates with a mind for business increasingly took management degrees, and many good physics and maths graduates sought their fortunes in the City of London.

From the 1980s, as national funding slowed, the University and UMIST had learned again how to look for local linkages and finance.

A Science Park was developed, but with limited success; it was not easy to make a Silicon Valley. The real resource was the steadily increasing demand for university education, which had become a *sine qua non* for almost all careers. Manchester, like other northern cities, came to rely heavily on university students, especially arts students, as major consumers of the leisure culture for which much of the city centre has been rebuilt.

In 2002, after a decade when UMIST and the Victoria University had loosened their ties of a century, the two institutions decided to merge. The old arrangement had always been confusing, especially for foreign applicants; the management and business schools had already merged, and some staff regretted the failure of 'rationalisation' between the two sets of engineering programmes. UMIST had slipped in the national research rankings, and life was getting harder for technical institutes with little bio-medicine or arts programmes. 'Project Unity' quickly gained general support.

The new University of Manchester, established in 2004, has four Faculties: three for the sciences, technologies and medicine, and one for everything else. Physical Sciences and Engineering brought together most of UMIST with much of the former Faculty of Sciences and Engineering in the Victoria University. The Medical School is now joined with Pharmacy and Psychology, as well as Dentistry and Nursing. And the Faculty of Life Sciences, which kept its recent independence, now includes the big new biomedical facility which the Wellcome Trust funded for UMIST (plus our Centre for the History of Science, Technology and Medicine, with its own Wellcome Unit).

Several of the new university's leaders have been bought in from outside, promising more professional management and more Nobel prize-winners. 'Step-changes' are intended, to make Manchester the premier university of the North, to rank with Oxbridge and the best of London. Our history may give ground for hope, for, as we have seen, that was indeed Manchester's position for about a century. Its leaders had drawn skilfully on Scottish, German and American patterns in education and science, adapting them to the city and region.

Again we have an historic opportunity, with great investments and great hopes, but it is still too early to judge how the new university will carry forward the character of this heritage and locality. Watch this space.

Notes

Images reproduced by courtesy of The University Librarian and Director, The John Rylands University Library, The University of Manchester.

1. This article was initially prepared for Clive Field and John Pickstone

(eds.), *A centre of intelligence* (Manchester, 1988; rev. 1992); a version is available on the CHSTM web site: www.manchester.ac.uk/chstm. It is thus biased towards the history of academic institution, especially the Owens College and the Victoria University. For more on primary sources see John V. Pickstone, 'Some Manchester sources for the history of science, technology and medicine, with special reference to The John Rylands University Library of Manchester', *Bulletin of the John Rylands University Library of Manchester* [hereafter *BJRULM*], 71:2 (1989), pp. 141–57. For recent work on medicine in Manchester, see the special volume of *BJRULM*, 'Medical history in Manchester: health and healing in an industrial city, 1750–2000' edited by Stella Butler and John Pickstone, vol. 87:1 (forthcoming, 2007). For brief but illustrated articles on sciences and medicine in Manchester University before and after 1905, see Jane Havell and Brian Pullan (eds.), *A portrait of the University of Manchester* (London, 2006). The standard work on nineteenth-century Manchester science remains Robert H. Kargon, *Science in Victorian Manchester* (Manchester, 1977). For wider histories and references see David Edgerton and John V. Pickstone, 'Science, technology and medicine in the United Kingdom, 1750–2000,' in David N. Livingstone and Ronald L. Numbers (eds.), *The Cambridge history of science*, viii: *Modern science in national and international contexts* (Cambridge, forthcoming 2007).
2. Peter J. French, *John Dee: the world of an Elizabethan magus* (1972); Benjamin Woolley, *The Queen's conjuror: the science and magic of Dr Dee* (London, 2001). For another Elizabethan mathematician with local connections, see Frances Willmoth, *Sir Jonas Moore, practical mathematics and Restoration science* (Woodbridge, 1993). See also Stephen Bowd, 'In the labyrinth: John Dee and Reformation Manchester', *Manchester Region History Review* [hereafter *MRHR*], 19 (forthcoming, 2008).
3. See the references in Chapman's article, including Charles Webster, 'Richard Towneley, 1629–1707, and the Towneley Group', *Transactions of the Historic Society of Lancashire and Cheshire*, 118 (1967), pp. 51–76.
4. Christopher Booth has published widely on this group: see B. C. Corner and C. C. Booth, *Chain of friendship: selected letters of John Fothergill, 1735–1780* (Cambridge, Mass; 1971); and several essays in Christopher C. Booth, *A physician reflects: Hermann Boerhaave and other essays* (London, 2003).
5. For introductions and references, see articles in the (newly revised) *Dictionary of National Biography* [hereafter *DNB*] and the *Dictionary of Scientific Biography*; and Nicholaas Rupke, *Richard Owen: Victorian naturalist* (New Haven, 1994).
6. The classic article on the Lit and Phil is Arnold Thackray, 'Natural knowledge in cultural context: the Manchester model', *American Historical Review*, 79 (1974), pp. 672–709.
7. See Jean Raymond and John Pickstone, 'The natural sciences and the learning of the English Unitarians: an exploration of the roles of Manchester

College,' in Barbara Smith (ed.), *Truth, liberty and religion: essays celebrating two hundred years of Manchester College* (Oxford, 1986).

8. For the hospitals and context see John V. Pickstone, *Medicine and industrial society: a history of hospital development in Manchester and its region, 1752–1948* (Manchester, 1985); and John V. Pickstone and Stella V. F. Butler, 'The politics of medicine in the early industrial city: a study of hospital reform and medical relief in late-eighteenth century Manchester', *Medical History*, 28 (1984), pp. 227–49; and John V. Pickstone, 'Thomas Percival and the production of medical ethics,' in R. Baker, D. Porter and R. Porter (eds.), *The codification of medical morality* (Amsterdam, 1993), pp. 161–78.

9. Elizabeth Patterson, *John Dalton and the atomic theory: the biography of a natural philosopher* (New York, 1970); Arnold Thackray, *John Dalton: critical assessments of his life and science* (Cambridge, Mass; 1972); and see the recent articles by Rajkumari Williamson Jones, 'John Dalton's summer holiday of 1840', *MRHR*, 16 (2002–3), pp. 13–21; and (with Terry Wyke), 'Representations and rememberances of scientists in Manchester', *MRHR*, 17:1, 2004, pp. 10–17.

10. R. F. Bud, 'The Royal Manchester Institution', in D. S. L. Cardwell (ed.), *Artisan to graduate* (Manchester, 1974), pp. 119–33, and Howard M. Wach, 'The condition of the middle-classes: culture and society in Manchester, 1815–1850' (unpub. PhD, Brandeis University, 1987).

11. Mabel Tylecote, *The Mechanics' Institutes of Lancashire and Yorkshire before 1851* (Manchester, 1957). See the article by Richard Hills below, and his *Life and inventions of Richard Roberts, 1789–1864* (Ashbourne, 2002).

12. For a tour of scientific Manchester c. 1840 see *Manchester as it is* (London, 1839; repr. Manchester, 1971), chap. 7.

13. For perspectives on science and technology in this analytical age, see John V. Pickstone, *Ways of knowing: a new history of science, technology and medicine* (Manchester, 2000), esp. chap. 4. For chemistry, see the six articles by W. V. Farrar, Kathleen R. Farrar and E. L. Scott on the Henrys of Manchester, in *Ambix*, 20–4 (1973–7); and Wilfred V. Farrar, *Chemistry and the chemical industry in the nineteenth century: the Henrys of Manchester and other studies* (eds.), Richard L. Hills and W. H. Brock (Aldershot, 1997). On medicine, fever and social transformation, see John V. Pickstone, 'Ferriar's fever to Kay's cholera', *History of Science*, 22 (1984) pp. 401–19; and 'Dearth, dirt and fever epidemics; rewriting the history of British public health, 1780–1850,' in Terence Ranger and Paul Slack (eds.), *Epidemics and ideas: essays on the historical perception of pestilence* (Cambridge, 1992), pp. 125–48; Mary Poovey, *Making a social body: British cultural formation, 1830–1864* (Chicago, 1995).

14. Thomas S. Ashton, *Economic and social investigations in Manchester, 1833–1933* (London, 1934; repr. Brighton, 1977); M. J. Cullen, *The statistical movement in early Victorian Britain* (Brighton, 1975); D. Elesh, 'The Manchester Statistical Society: a case study of a discontinuity in the history of

empirical social research', *Journal of the History of the Behavioural Sciences*, 8 (1972), pp. 280–301, 407–417.
15. Kargon discusses the industrial chemists at length, especially their public health work; and see fn. 13.
16. See Nathan Rosenberg and Walter Vincenti, *The Britannia Tubular Bridge: the generation and diffusion of technological knowledge* (Cambridge, Mass; 1978).
17. Peter Ellerton Russell, 'John Frederic La Trobe-Bateman, F.R.S., water engineer 1810–1889' (unpub. MSc, University of Manchester, 1980).
18. D. S. L. Cardwell, *James Joule: a biography* (Manchester, 1989); and Crosbie Smith and M. Norton Wise, *Energy and empire: a biographical study of Lord Kelvin* (Cambridge, 1989).
19. See Raymond and Pickstone, 'Natural sciences and the learning of the English Unitarians'.
20. Joseph Thomson, *The Owens College: its foundation and growth* (Manchester, 1886); E. Fiddes, *Chapters in the history of Owens College and of Manchester University* (Manchester, 1937); W. H. B. Charlton, *Portrait of a university* (Manchester, 1951). For a recent account of Manchester along with other provincial universities, see Samuel J. M. M. Alberti, 'Civic cultures and civic colleges in Victorian England,' in Martin Daunton (ed.), *The organisation of knowledge in Victorian Britain* (Oxford, 2005), pp. 337–56.
21. On Jevons, see Margaret Schabas, *A world ruled by number: William Stanley Jevons and the rise of mathematical economics* (Princeton, 1990); and Phillip Mirowski, *More heat than light* (Cambridge, 1989). The children's author and conservationist Beatrix Potter was Roscoe's niece.
22. Robert Bud and Gerrylynn Roberts, *Science versus practice: chemistry in Victorian Britain* (Manchester, 1984).
23. M. R. Fox, *Dye-makers of Great Britain, 1856–1976: a history of chemists, companies, products and changes* (Manchester, 1987).
24. Henry Roscoe, 'Original research as a means of education', in *Essays and addresses by the professors and lecturers of the Owens College, Manchester* (London, 1874); and John V. Pickstone, 'Science in nineteenth-century England: plural configurations and singular politics', in Daunton, *The organisation of knowledge*, pp 337–56.
25. J. D. Jackson, 'Osborne Reynolds: scientist, engineer and pioneer', *Proceedings of the Royal Society*, Series A, 451 (1995), pp. 49–86.
26. *DNB* article on Balfour Stewart; Graeme Gooday, 'Precision measurement and the genesis of teaching laboratories in Victorian Britain' (unpub. PhD, University of Kent, 1989); and 'Precision measurement and the genesis of teaching laboratories in Victorian Britain', *British Journal for the History of Science*, 23 (1990), pp 25–51; Peter Davies, 'Sir Arthur Schuster, 1851–1934' (unpub. PhD, University of Manchester, 1983).
27. See his article in this volume, and his forthcoming history of the Manchester Museum.

28. For more on biology in Manchester, see Alison Kraft and Samuel J. M. M. Alberti, '"Equal though different": laboratories, museums and the institutional development of biology in late-Victorian northern England', *Studies in History and Philosophy of Biological and Biomedical Science*, 34 (2003), pp. 203–26; Alison Kraft, 'Building Manchester biology, 1851–1963' (unpub. PhD, University of Manchester, 2000); W. Alan Charlton and Elizabeth G. Cutter, *135 years of botany at Manchester* (Manchester, 1986).
29. See Cardwell, *Artisan to graduate*.
30. See the article on Delépine in Willis J. Elwood and Felicité A. Tuxford (eds.), *Some Manchester doctors: a biographical collection to mark the 150th anniversary of the Manchester Medical Society, 1834–1984* (Manchester, 1984); and for more on various Manchester doctors and their versions of germs see Michael Worboys, *Spreading germs: disease theories and medical practice in Britain, 1865–1900* (Cambridge, 2000).
31. See Pickstone, *Medicine in industrial society*; and Helen K. Valier, 'The politics of scientific medicine in Manchester, 1900–1960' (unpub. PhD, University of Manchester, 2002).
32. Brian Robinson, *The history of pharmaceutical education in Manchester* (Manchester, 1986).
33. On nineteenth-century funding and education, see D. S. L. Cardwell, *The organisation of science in England* (London, 1957).
34. See Davies, 'Sir Arthur Schuster'. On technical education, Anna Guagnini, 'The fashioning of higher technical education in Britain: the case of Manchester, 1851–1914', in H. F. Gospel (ed.), *Industrial change and technological innovation: a comparative and historical study* (London, 1991).
35. Stella V. F. Butler, 'A transformation in training: the formation of university medical faculties in Manchester, Leeds, and Liverpool, 1870–84', *Medical History*, 30 (1986), pp. 115–32.
36. For Harden and many other eminent Mancunian scientists, see the *Biographical memoirs of the Fellows of the Royal Society of London*.
37. Norman Rose, *Chaim Weizmann: a biography* (New York, 1986).
38. On sanitation and environmental politics see Alan Wilson, 'Technology and municipal decision-making: sanitary systems in Manchester, 1868–1910' (unpub. PhD, University of Manchester, 1990) and Harold L. Platt, *Shock cities: the environmental transformation and reform of Manchester and Chicago* (Chicago, 2004). Look out for forthcoming work by Harriet Ritvo on Manchester's water, and Daniel Schneider on activated sludge.
39. David Wilson, *Rutherford: simple genius* (Cambridge, Mass; 1983).
40. Ray Monk, *Ludwig Wittgenstein: the duty of genius* (London, 1990).
41. See the *Biographical memoirs of the Fellows of the Royal Society*, esp. for Bohr; and the biographical introduction to John Slater (ed.), *The collected works of Samuel Alexander* (London, 2000); Ruth Hall, *Marie Stopes* (London, 1978).

42. For wonderful accounts of Manchester's Edwardian culture, from below and from above, and for the arts and for the university and industry, see Neville Cardus, *Autobiography* (London, 1947) and Katharine Chorley, *Manchester made them* (London, 1950).
43. Alison Kraft, 'Pragmatism, patronage and politics in English biology: the rise and fall of economic biology, 1904–1920', *Journal of the History of Biology*, 37 (2004), pp 213–58.
44. See *Biographical Memoirs of the Fellows of the Royal Society*.
45. Willis Jackson came from a working-class family in Burnley: his baptismal name was Willie. For him, Freddie Williams, Tom Kilburn and more, see T. E. Broadbent, *Electrical engineering at Manchester University* (Manchester, 1998).
46. See Tim Cooper's chapter and his thesis, 'The science-industry relationship: the case of electrical engineering in Manchester 1914–1960' (unpub. PhD, University of Manchester, 2003); Chris Niblett, 'Images of progress: three episodes in the development of research policy in the UK electrical engineering industry', (unpub. PhD, University of Manchester, 1980); G. Tweedale and M. Sawai, 'Industrial research in Osaka and the north-west UK from the 1920s to the 1960s', in D. Farnie *et al.* (eds.), *Region and strategy in Britain and Japan: business in Lancashire and Kansai, 1890–1990* (London, 1999), pp. 252–99. And see Cooper's list of firms doing research in the Manchester region.
47. For the MRI and science, see Valier, 'The politics of scientific medicine', fn. 31 above. For the interwar period more generally see John V. Pickstone, 'Manchester's history and Manchester's medicine', *British Medical Journal*, 295 (1987), pp. 1604–8; and Sarah Barnes, 'England's civic universities and the triumph of the Oxbridge ideal', *History of Education Quarterly*, 36 (1996), pp. 272–305.
48. See the Special Number of the *IEEE Annals of Computing*, 115 (1993); Alan Hodges, *Alan Turing: the enigma of intelligence* (London, 1983); Simon Lavington, *A history of Manchester computers* (Manchester, 1976).
49. See Brian Pullan and Michelle Abendstern, *A history of the University of Manchester, 1951–1973* (Manchester, 2000); and *A history of the University of Manchester, 1973–90* (Manchester, 2004).
50. Duncan Wilson (CHSTM) is working on a history of this key merger, to be published as a book and in the form of a scholarly article by Duncan Wilson and Gael Lancelot.
51. On Ferranti, ICL etc. see: J. F. Wilson, *Ferranti and the British electrical industry, 1864–1930* (Manchester, 1988); Martin Campbell-Kelly, *ICL: a business and technical history* (Oxford, 1989).

Under a Lancashire heaven: William Crabtree, Jeremiah Horrocks and their circle, and the origins of research astronomy in seventeenth-century England

Allan Chapman

The scientific revolution in continental Europe was, in so many ways, the product of institutions. The extensive analysis of astronomical theories and tables, and their errors, which led Nicholas Copernicus to propose in 1543 that the earth moved around the sun, was accomplished through the time and resources made available to him from the revenues of the Polish Cathedral where he was a canon. Tycho Brahe's observation of the northern heavens between 1572 and 1597, which released a flood of new data into the scientific community, was performed on the strength of generous official grants from the Danish crown; while both Galileo and Johannes Kepler, in addition to holding university professorships, also benefited from the support of powerful monarchs. Galileo, after rising to fame in 1610, became a scientific courtier to the Medici family in his native Tuscany; while Kepler inherited Tycho's last appointment – though not his lavish revenues – as Mathematician to the Holy Roman Emperor Rudolf II.[1]

All of these men, and a good number besides, had come to wrestle with one of the most intractable intellectual problems of the Renaissance: why did the planets not move as the astronomical tables said that they should? Why was it that, when one took the historic observations and tables of the Greeks such as Hipparchus and Ptolemy, of medieval Europeans such as the Castilian astronomers who produced the Alfonsine tables, and Arabs like Al-Zarqali, or of the fifteenth-century German Regiomontanus, they all failed to predict the motions of the heavens with accuracy? Of course, all these astronomers took it as axiomatic that the sun, moon, planets, and stars rotated around an earth that was fixed in the centre of the universe, and believed that one could explain the apparent speeding up and slowing down of certain planets, such as Mars, by making adjustments to the geometrical models which they believed could be used to account for the complex motion of the heavens.

The idea that the earth moved around the sun, and that the earth and planets might well move in elliptical and not circular orbits, was an act of immense imaginative power. The notion that the earth might be spinning in space flew in the face of reason and common sense, for how did we avoid being flung into space? On the other hand, these seemingly absurd ideas did make it much easier to model the movements of the planets, so that observed astronomical phenomena agreed with geocentric mathematical predictions.

Behind the new evidence was a radically new concept of the power of sense-knowledge, including the very precise data revealed to us by the telescope and recently invented precision angle-measuring instruments. In the 'Preface' to his *Micrographia*, in 1665, Robert Hooke was to refer to these instruments as 'artificial Organs', in so far as they presented to the human senses an abundance of previously unavailable information.

By 1630, the nature of the heavens, the possibility of the sun being at the centre of the solar system, and the interpretation of the new telescopic discoveries, such as Galileo's discovery of the moons of Jupiter or the spots on the sun, were being discussed across learned Europe. In the British Isles, people read the works of Galileo and Kepler, and bought telescopes – often made in Holland or Paris – with which to look at the planets for themselves. Yet the British response to the continental discoveries was generally conservative; one of admiration, wonder and gradual acceptance. For even by the 1630s no Englishman (except the reclusive, unpublished and then deceased Thomas Harriott) had made major independent discoveries in the New Astronomy, whether in London, Oxford, Cambridge, or at the artistically cultured court of King Charles I.[2] At the same time as Italy was giving expression to the genius of Galileo, England was witnessing the flourishing of Shakespeare.

At Gresham College, London, in the early seventeenth century, the Copernicans Edmund Gunter and Henry Gellibrand were making major experimental advances in our understanding of the nature of terrestrial magnetism (in the wake of William Gilbert's *De Magnete* of 1600), and they even saw the action of the earth's magnetic field as congruent with its spinning in space, but the next advances in astronomy, after those of Galileo, Kepler, and Pierre Gassendi, unfolded in the most unexpected of places: rural Lancashire, and across the Pennines in West Yorkshire. This was a region of England, moreover, which was thinly populated, relatively poor, and lacking in major educational resources. One of the very last places, indeed, where one might have expected to find a group of men who were actively wrestling with the great continental discoveries and making advances beyond them. For the astronomical achievements of Florence, Prague and Paris were being examined and developed in Salford, Liverpool, and in the villages of Much Hoole near Preston, and Middleton just outside Leeds.

Why the English astronomical Renaissance should have happened in this north-western environment is not easy to explain. There were no especial lines of patronage in the region that were fundamentally different from what one might have expected to find in Norfolk or Devonshire, let alone what could have been possible in London, Oxford or Cambridge. Intellectually-inclined county squires, small merchants, and clergy were spread across the country, though their interests tended more to literature, theology or the arts rather than to science. Yes, Liverpool was a port, but in 1630 it was still very much a regional port serving the Irish Sea and the coasting trade and would not necessarily have been stimulated by the astronomical needs of global navigation. In that respect, London, Bristol or Plymouth would have provided more fertile astronomical soil. There seems no obvious set of social or economic circumstances that can be invoked to explain why the European Astronomical Renaissance took root in rural Lancashire and West Yorkshire beyond the contingent presence of a small group of individuals who happened to be interested in the subject. Historical reality is deeply idiosyncratic.

The most intellectually significant of the local astronomers was Jeremiah Horrocks (1618/19–1641), the son of a Toxteth, Liverpool, yeoman farmer with family connections to the local watch-making

Starr Cottages, Broughton Spout, home of William Crabtree, the Salford astronomer (Higher Broughton, Salford). A copy of the original formerly in the possession of Tom Fern, late of Salford Astronomical Society

Starr Cottages, home of William Crabtree. Water colour painting by P. Roughan, 1991

trade. Horrocks's brother Jonas also had astronomical interests and probably owned a telescope, though we know nothing of Jonas beyond a few passing remarks by Jeremiah. Then there was William Crabtree (1610–1644), a clothier, or cloth-dealer, of Salford, who was already established as a serious astronomer by 1636 and was a formative influence on the young Horrocks.[3] And there was also William Gascoigne (1612–1644) who seems to have been descended from an old Catholic landed family at Middleton near Leeds, and who claimed in one of his surviving letters to have received part of his education at Oxford.[4]

In spite of the relative isolation of rural Lancashire at this time, however, it is clear that many middle-class families – yeoman farmers and minor gentry, and urban merchants and manufacturers – made sure that their sons were well educated. The Horrocks family and the Aspinwalls (Jeremiah's mother's side) were comfortably-off yeoman farmers of Bolton and Toxteth, with business interests in watchmaking. Indeed, as Peter Aughton tells us, these two families sent 25 of their sons to Oxford and Cambridge between 1542 and 1650: the Horrocks family favouring Cambridge and the Aspinwalls Oxford.[5] Jeremiah, the Cambridge graduate and starving curate of popular legend was, in fact, the scion of two established middle-class families. And while we know less about the Salford Crabtrees, they were clearly of established standing in the region.[6] On 8 July 1619, for instance, when William the astronomer was nine years old, another 'William Crabtree of Broughton in the County of Lancaster, Yeoman' signed a deed relating to land in the Levenshulme district of Manchester.[7] And in Chetham's College Library is a 'Book of Rates for the County

Pallatine of Lancaster' signed by one W. Crabtree.[8] This document is undated, but appears to have been produced after August 1644 when – from a surviving will and a burial register – *our* William Crabtree seems to have died.[9] Our William, the astronomer, married Elizabeth Pendleton of Pendleton, Salford, in September 1633; he was sufficiently well off to have acquired a sound reading knowledge of Latin and to collect expensive imported books, and to devote a significant amount of his time to the financially unrewarding pursuit of planetary motion. He probably acquired the books through his contacts in the textile trade which in those days operated through London and Flanders.

William Crabtree had attended the 'Manchester School', which would probably have been the grammar school attached to the Collegiate Church (now Manchester Cathedral); he does not seem to have been a university graduate. Horrocks and Gascoigne had attended university (Horrocks had matriculated at Emmanuel College, Cambridge, where he resided between 1632 and 1636), and Gascoigne seems to have spent some time in Oxford and to have travelled abroad, as will be shown presently, though neither had taken degrees. So how did Crabtree and Horrocks meet? It is possible that Horrocks's Emmanuel contemporary, John Worthington of Manchester, brought the two men together in 1635 or 1636.[10] They were certainly corresponding about abstruse astronomical matters by 21 June 1636, and would continue to do so up to Horrocks's sudden death in January 1641.[11]

Indeed, an intense intellectual passion runs through the writings of these men. As Crabtree, the Lancashire clothier and small businessman, confessed to Gascoigne in a letter of 30 October 1640, '[o]f all Desires the Desire of Knowledge is most vehement, most impatient: and of all Kinds of Knowledge, this of the Mathematicks affects the Mind with most intense Agitations'.[12] None of these men would have encountered the New Astronomy of the European Renaissance on a formal curricular basis at school or university, but all three would have received a good grounding in Latin, and thus had access to the writings of Tycho, Galileo, Kepler, Gassendi, which they mention in their surviving correspondence, along with the works of lesser figures such as Philip Lansberg, Ismael Bouilleau and others. We also know that Crabtree was in correspondence with Samuel Foster, Professor of Astronomy at Gresham College, London (and a 1616 alumnus of Emmanuel College).[13]

It is not unlikely that this correspondence with Foster, who was elected Gresham Professor of Astronomy in 1636, had begun with a personal meeting. As the lectures of the Gresham College Professors were statutorily open to the public, it is possible that Crabtree could have attended them when visiting the City on business.[14] After all, Sir

Thomas Gresham, the founder of the College and the endower of its seven professorships in 1597, had himself been a mercer, and the Mercers' Company, with its historic links with the textile trade was, and still remains, the City Livery Company most concerned with the College's academic activities. Foster, moreover, was an established author on practical mathematics, having published a book on the use of the quadrant in 1624, and another on the geometrical construction of sundials in 1638.

Quite independently of Horrocks and Crabtree, and before he first came to know them (probably in 1640), William Gascoigne was a correspondent of the Revd William Oughtred, vicar of Albury, Surrey, and one of the most influential English mathematicians of the first half of the seventeenth century. And in 1841, Stephen J. Rigaud published some of this correspondence. Yet how the Yorkshire Roman Catholic gentleman struck up this correspondence with the very Anglican Surrey parson we have no idea, though one suspects the existence of a wide correspondence network which cut across regional, social, and sectarian lines.[15]

Indeed, these three figures, Horrocks, Crabtree and Gascoigne, very clearly had correspondence with and connections to a wider diaspora of learning not merely in London or in the universities, but in the north-west region and did not in any way operate in isolation. All were connected with the brothers Charles (Snr) and Christopher Towneley of Towneley Hall near Burnley, Lancashire, and to a lesser extent with the Sherburnes of nearby Stonyhurst. Both the Towneleys and the Sherburnes were old Roman Catholic gentry families, largely frozen out of public life because of their faith, yet still generous patrons of learning in their native Lancashire. It is almost certain that they were connected with the Gascoignes, a family of the same social rank and Catholic faith, living just a few miles to the east on the Yorkshire side of the Pennines. The Towneleys in particular were active in science both as practitioners and patrons, through at least two generations, as was demonstrated in Dr Charles Webster's major paper in 1966.[16] In the early 1660s, for instance, the Halifax, Yorkshire, physician Dr Henry Power was conducting barometric experiments at the top and bottom of Pendle Hill, Lancashire, with Charles Towneley (Jnr), along with investigations into the 'fiery' and other 'damps' found in coalmines and other deep pits in the vicinity.[17]

Richard Towneley, son of Charles Towneley (Snr), elder brother of Charles (Jnr), and nephew of Christopher, became in his generation a friend and correspondent of the Revd John Flamsteed, the first Astronomer Royal, in the 1670s. When, aged 77, Richard drew up his will, he specifically mentioned his scientific instruments, 'many of them being valuable and hard to be gott'.[18] The collection included the surviving instruments and papers of Horrocks, Crabtree and Gascoigne, which Richard had made available when the young Flamsteed visited Towneley Hall in 1672. Flamsteed was to play a key part in passing on their achievements to the Royal Society, thus to the wider realm of learning.

But Richard Towneley enjoyed a considerable reputation in his own right – at least as far as was compatible with his status as a Roman Catholic who felt it wisest to remain in his Lancashire fastness. For in addition to the acquisition of the collection of scientific instruments, Richard had undertaken pioneering researches in meteorology, keeping detailed weather records at Towneley Hall, doing work on air pressure which helped to inspire Boyle's Law, and studying the geology of Pendle Hill. In 1675, moreover, Towneley proposed designs for a new type of clock, for Flamsteed's two-second-beat regulator clocks with pendulums 13 feet long. This Towneley mechanism, realised in brass by Thomas Tompion, was an early form of 'dead beat' escapement, capable of considerable accuracy, though it demanded the highest skills in manufacture. Flamsteed's clocks were paid for by the now wealthy Sir Jonas Moore, FRS, youthful protégé of Gascoigne, friend of the King, patron of Flamsteed, horological experimenter in his own right. Moore was born and raised not far from Towneley Hall, and became a leading apostle of the North Country astronomers to the Royal Society after 1660.[19]

Clearly, the North Country astronomers were not simply inspired local geniuses, working in social isolation, and struggling against poverty and overwhelming odds, as the fashionable Victorian depiction implied. But we have yet to explain how Crabtree and Horrocks met Gascoigne.

In his 'Life' of Christopher Towneley, in the Appendix to his *Sphere of Marcus Manilius* (1675), Sir Edward Sherburne claimed that it had been Christopher who brought *all three* men, Horrocks, Crabtree and Gascoigne, together, which may well have been the case. Yet how these Towneley connections were made originally is not itself clear, though as the Towneleys were already established patrons of learning in Lancashire by 1640, Christopher may well have approached Crabtree in the same way as he brought the young Jonas Moore, Jeremy Shakerley, Henry Power and others into their family orbit.

Evidence suggests that Crabtree was the principal 'intelligencer' who linked Horrocks, Gascoigne and perhaps Towneley; he actively corresponded with all three men and no other letters survive from the group. It also appears unlikely that Crabtree and Horrocks were in correspondence with Gascoigne before 1640, for while John Flamsteed found and printed a Gascoigne observation of 10 December 1638, this was in a private note of Gascoigne's and not in a letter.[20] Gascoigne's name is absent in all of the documents relating to the 24 November 1639 transit of Venus.[21] Horrocks does not mention him in the crucial letter to Crabtree of 26 October 1639, where he asks Crabtree to contact other potential observers;[22] nor does Gascoigne's name appear in Horrocks's *Venus in sole visa* in which he describes his own and Crabtree's successful observations of the same transit.

By 7 August 1640, however, Crabtree and Gascoigne were corresponding on extensive and familiar terms, though the Latin tag above Crabtree's signature attached to this letter – *de facie ignotus* (you do not know my face) – tells us that they had not yet met. Even so, in this same letter a cross-Pennine trip to Gascoigne at Middleton was being planned, no doubt before the winter set in, for Crabtree expressed his wish to bring over Horrocks 'with Mr. Towneley and my self, to see Yorkshire and you'.[23] And from this time onwards the correspondence grew apace: about the making of specific observations, the design of Gascoigne's instruments, and the discoveries of Horrocks; Gascoigne's telescopic sights and micrometer were discussed in detail. Of special significance was their discussion of Horrocks's work on the ellipticity of the lunar orbit, and the testing of his Lunar Theory by regular critical observation. Gascoigne too was a keen observer, for a large number of his lunar and planetary diameter observations survive, having been printed by John Flamsteed in 1725 and by Rigaud in 1841. One also senses that, following the death of Horrocks, William Gascoigne became Crabtree's foremost astronomical correspondent down to 1643.

Yet what did Horrocks, Crabtree and Gascoigne do that was so original? They were certainly not the first English Copernicans, nor were they the first to read the works of Galileo and Kepler. Thomas Harriott and Sir William Lower had probably discovered sunspots as

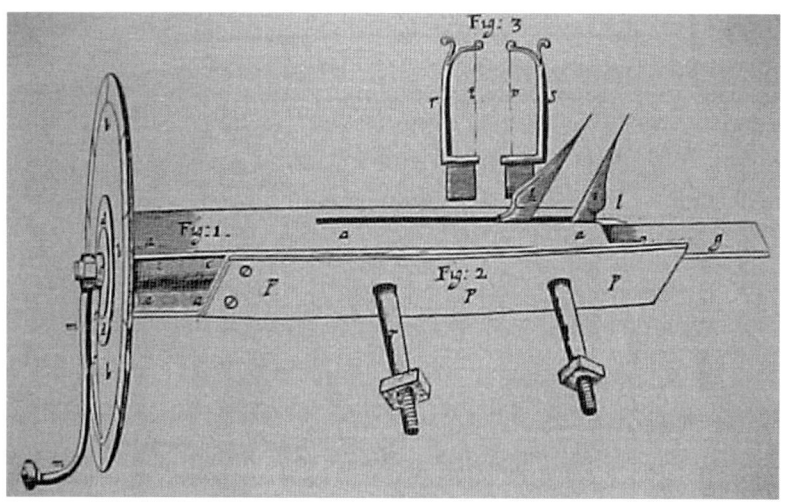

William Gascoigne's micrometer, developed by Richard Towneley (from *Phil. Trans.*, 2, 1667; and reproduced in A. Chapman, *Dividing the circle: the development of critical angular measurement in astronomy, 1500–1850*)

early as December 1610, some months before Galileo, though they did not seem to fully grasp their significance and failed to publish their findings. The real importance of Horrocks and his circle lay in their attitude towards the nature of new knowledge and the emphasis which they placed upon original observations made with modern instruments (as opposed to a relatively uncritical acceptance of the achievements of the past). They were confident that they, and fellow 'moderns', belonged to a new epoch in the history of astronomical discovery. Their new trust in instrument-derived sense-knowledge is clearly manifested in Horrocks's and Crabtree's correspondence after 1636 where they express disillusionment with the errors present in the standard published tables – Horrocks had an especial animus against the *Tabulae motuum* of Philip Lansberg – and in their constant concern with observing, measuring and recording. It is also displayed in William Gascoigne's invention, around 1639, of the telescopic eyepiece micrometer, whereby very tiny angles could be measured to a critical level of accuracy through the telescope.[24]

But most of all, Horrocks's, Crabtree's and Gascoigne's scientific confidence derived from the telescope itself. Galileo's 'perspective cylinder' of 1610 showed that the telescopic universe was a very different place from the universe seen by the naked eye. To the naked eye the planets were but moving stars, whereas through the telescope they appeared to be spheres and worlds in their own right. Jupiter even had four satellites going around it, thereby nonplussing those conservative astronomers who said that the earth was the only point of rotation in the universe because it was at the centre. Venus showed phases which strongly suggested that it turned around the sun and not the earth; while even the sun itself, far from being serene and changeless as the ancients had taught, was revealed by the telescope to rotate on its axis every 28 days and was sometimes blotted with sunspots. And perhaps it was the moon which most forcibly undermined classical astronomical authority: through the telescope, the lunar surface showed a complex geography of craters, 'seas', bays and mountains that reminded one of maps of the earth's surface. Was the moon a world like our own, with 'Selenite' (from the Greek moon goddess Selene) inhabitants?

Horrocks, Crabtree and Gascoigne had invented none of these ideas, which had spilled across Europe after Galileo's monumental *Siderius nuncius* ('Starry messenger') of 1610 had made him famous across the world of learning. But the Galilean discoveries made people realise that if they could build yet more powerful telescopes and even more delicate measuring instruments, then they might discover still more celestial wonders. This sense of new possibilities suffuses the correspondence of Crabtree and Horrocks during the late 1630s, and in the letter of 7 August 1640, sent to William Gascoigne, Crabtree describes his own telescopic observation of some recent sunspots and uses them as evidence of the fact that the sun possesses an independent axial rotation. In this same letter, Crabtree opened up discussion about the possible nature of these spots which he likened to 'the Smoak of some pitchy Fire, which is in one part very gross, an in another more rare' in their dense umbral and lighter penumbral regions. Each of these men owned at least one telescope – Horrocks specifically mentions possessing two.

In addition to the telescopic discoveries, the new celestial mechanics of Johannes Kepler had a profound influence upon the North Country astronomers, and from a surviving list of 31 astronomical books which Jeremiah Horrocks made in 1635, no less than four titles were works by Kepler.[25] Crabtree, moreover, seems to have owned an even larger astronomical library. It is interesting to note, however, that none of the books in Horrocks's 1635 list was by Englishmen or in the English language. Some of Horrocks's books were classical texts while two authors, Albategnius and Alfraganus, were medieval Arabic astronomers whose works were known in the west via sixteenth-century Latin translations. Most were by modern European writers and all were in the Latin language. By 1637, however, Horrocks was using at least one English source when, in a letter to Crabtree of 23 November 1637, he mentioned using an eye-parallax correction for his new cross staff based on the data provided by Edward Wright: probably Wright's *Errors of navigation* (1599).[26] Indeed, it is surprising that more English authors do not appear, writing either in their mother tongue or in Latin, for by 1635 English authors such as Robert Recorde, Leonard and Thomas Digges, John Dee, William Gilbert, Samuel Foster, Henry Briggs, Henry Gellibrand and many others were freely available. Much of these English writings, however, were concerned with geomagnetism, pure and applied mathematics and practical navigation, and one looks in vain for serious, up-to-date astronomical research.

Kepler's initial significance lay in his 1608 announcement that the planet Mars, instead of moving in a circular orbit, moved in an ellipse. Then in 1619, Kepler amplified his original discovery by describing an exact mathematical ratio in accordance with which Mars, and by extension the other planets, sped up and slowed down as they orbited. And of course, it was around the sun that Kepler argued the planets orbited, and not the earth. This new Keplerian celestial mechanics provided astronomy with its first exact mathematical laws; for in Kepler's eyes, the planets did not move because they were attached to great transparent spheres which rotated about the earth, they moved in empty space under the control of some invisible yet mathematically exact force which he believed emanated from the sun. Jeremiah

Horrocks was the first scientist in Britain and in Europe to take the concept of orbital ellipticity and develop it further. He was also fascinated by Kepler's idea of 'Magneticall and Sympathetical Rays' emanating from the sun and driving along the planets which serves to remind us how powerful a concept magnetism was in the astronomical thought of this period. Indeed, in 1638, Horrocks's slightly older contemporary the Revd John Wilkins was to write a book on the possibly magnetic character of the earth's attractive force, and how one might overcome it, and journey to the moon in a 'Flying Charriot'.[27]

From a series of meticulous observations and measurements, Horrocks had come to conclude by 1638 that the moon also moved around the earth in an elliptical orbit, of which the earth occupied one of the geometrical foci. This breakthrough enabled him to begin to make sense of the very complex orbit which the moon describes around the earth when its changing position is measured against the positions of the fixed stars. This changing earth-moon distance, moreover, predicated that the angular diameter subtended by the moon, as observed from the earth, must vary across the lunar cycles, and if one could measure these changing angular sizes accurately, one could check the theory.

This is one reason why the telescope eyepiece micrometer which William Gascoigne invented around 1640 was to become so crucial; its capacity to measure celestial angles to within a few arc-seconds enabled one to apply stringent tests to a theory. For as the North Country astronomers came intuitively to understand, scientific knowledge was to be most effectively advanced when new discoveries enabled the framing of new hypotheses and where in turn, these hypotheses had to pass the acid test of scrupulous physical investigation and proof if they were to stand firm. It was not for nothing, therefore, that when Crabtree saw Gascoigne's early micrometer in 1640 he wanted one for Horrocks and himself and asked Gascoigne 'Could I purchase it with Travel, or procure it for Gold?'

Horrocks's work on the moon's motion, and his theory of how it moved in an elliptical orbit, would have been spectacular by any standards; but then, from his analyses of the orbital data, Horrocks drew a further conclusion. Namely, that the long axis – or apside line – through the lunar orbital ellipse does not remain in one place with reference to a given point in the starry background, but actually rotates back upon itself, or precesses, as the earth-moon-sun position constantly changes in space. These lunar orbital dynamics discovered by Horrocks would later prove invaluable to Sir Isaac Newton when he developed his theory of Universal Gravitation half a century later.

At the heart of Crabtree's, Horrocks's and Gascoigne's work was a great deal of practical, hands-on observation of the heavens, especially of the sun, moon and planets as they moved amongst the stars of the zodiac. And whilst these men used the basic geometry of the zodiac, with its 12 signs each encompassing 30° of sky, it should be emphasised that they displayed no sign of interest in astrology or astrological prediction. Indeed, in some of their letters and surviving writings they pour contempt upon astrology. They simply used the immemorial zodiac divisions to monitor the physical motions of the planets.

Measuring the exact distance of the sun, moon, and planets from each other, and from the First Point of Aries (Spring Equinox), was crucial for these North Country astronomers because that was the way in which they uncovered the errors in the standard astronomical tables and went on to draw original conclusions about the planetary motions. And for this, they used a variety of mathematical instruments, probably home-made, whereby they could read degrees and minutes of arc in the sky. For example, both Horrocks and Crabtree owned several types of 'Astronomical Radii' and at least one quadrant for measuring lateral and vertical angles in the sky, while in addition Crabtree seems to have timed observations with a clock.[28] With these instruments, they were able to construct new tables for the motions of the planets that were more accurate than the published ones. Indeed, Horrocks's and Crabtree's letters are full of these observations, measurements and comparisons with the standard published tables, and it is clear that they were consciously trying to obtain regular results that were more accurate and more reliable than those found in the tables.

We do not know whether William Gascoigne's invention of the micrometer was spurred on by what Crabtree and Horrocks were doing in Lancashire or whether the micrometer was the product of his own wrestlings with the same problems as they were dealing with. Either way, the micrometer came from the astronomer's need to be able to measure very small angles through the telescope, to see if the variations conformed to predicted Keplerian criteria, or else to note the positions of the planets from dim stars in the starry background. For while the telescope had been around as a scientific instrument since 1609, it had so far proved impossible for astronomers to devise a way whereby a pair of marker points – like the crosshairs in a modern telescopic rifle sight – could be inserted into the telescope's field of view to enable the observer to measure the relative positions of objects in that field.

Gascoigne made that crucial optical discovery – where to place a marker to make it stand out in a telescope field – at his father's house near Leeds in 1640 and then went on to develop it further. He next had the bright idea of inserting two marker points into the same field and controlling each with a delicate screw so that an astronomer looking at the moon, let us say, could carefully adjust one screw-controlled marker point to one side of the moon, and the other to the other side. Once this had been achieved, the astronomer could use his knowledge of the optical geometry of his telescope to calculate the angle subtended between the two marker points to a critical level of accuracy.[29]

It has already been mentioned that the surviving writing of Horrocks, Crabtree, Gascoigne and Towneley refer to several types of instruments, and to telescopes. Crabtree also owned a substantial quadrant with which he could measure vertical angles in the sky, and as he seems in addition to have undertaken local land survey and cartographic work (a map by him of Humphrey Booth's estate at Blackley, Manchester, survives from 1637), he probably owned various levels and surveyors' instruments as well.[30] The 'Astronomical Radii' which both Horrocks and Crabtree owned would have consisted of a 'T' configuration of wooden rods with sliding brass sights with which to measure either terrestrial or celestial angles, while Crabtree recorded timing a lunar eclipse by means of a clock as set out in his letter of 22 June 1642, published by Flamsteed in 1725.[31] So where could these instruments have come from?

Well, the telescopes had quite likely been made up by the astronomers themselves from lenses bought from spectacle makers. Commercially-made telescopes are referred to, though, as Horrocks on one occasion speaks of the quality of a 'half-crown telescope'. Clocks however would almost certainly have been purchased from professional manufacturers, being complex, high-status artefacts in 1640. While we do not know where Crabtree's clock came from, we do know that north-west England was a major centre for horological manufacture, primarily as a cottage industry.[32] The Horrocks and Aspinwall families of Toxteth, it will be recalled, earned part of their income from watch-making, while the Prescot area of west Lancashire near Liverpool developed a lively watch-making trade in the seventeenth century. While many Lancashire timepieces were sent out into the world complete, and with local makers' names on their faces, the local cottage industries also produced gears, pinions, and springs, and sometimes assembled the parts into working but uncased movements. These movements would then be exported to London on packhorses for finishing, casing and selling in the metropolis and perhaps to be exported overseas. Indeed, horological manufacture was clearly a thriving local trade, for 30 years later, in 1672, Robert Hooke recorded in his diary (16 August) having dealings with a 'Lancashire watchmaker's son about wheel cutting engine' for the accurate graduation of horological gear wheels. So while we know no more about Crabtree's clock beyond its bare existence and use for timing eclipses, it is not unlikely that he owned a timepiece which had been made relatively locally.

Watch-makers, of course, were the obvious persons to employ for the manufacture of any fine and delicate mechanisms, though rather annoyingly for us, neither Crabtree nor Towneley ever really specifies the precise trades of the 'Workmen' to whom they refer in their letters as the constructors of the whole or particular parts of their instruments. It is true that Gascoigne makes reference to a man who must have been a blacksmith, and on another occasion laments having no craftsman more skilful than a 'Country Joiner' to make a frame for a sextant, but this is all rather vague as to detail.[33] The two men clearly understood what was required and what was available and did not need to be more explicit in their letters. However, when Richard Towneley made an improved version of Gascoigne's micrometer to show to the Royal Society in 1667 he did specify engaging the services of 'an exact and ingenious Watch Maker of these Parts' in its construction.[34]

Access to skilled men must have been limited, though, for in one undated letter, Gascoigne lamented that while he had placed an order for 'an Iron Quadrant of Five Foot, which will give me the 1000th Part of One Degree', it would not be finished in time to observe the forthcoming eclipse as the craftsman in question was 'Throng' with other business for Mr Gascoigne senior, William's father. (William Derham, who seems to have owned this and other Crabtree and Gascoigne letters, selections of which he published in *Philosophical Transactions* in 1717, explained that 'Throng' was 'A Yorkshire Phrase for fully employed'.)

William Crabtree also commissioned craftsmen to make metal parts for him. On 30 October 1640, for instance, he informed Gascoigne that he planned in the near future to make a visit to 'Wigan, 12 Miles from hence [Salford] where much brass is cast' to see if it might be possible to commission the casting of a large quadrant (Derham also seems to have come to own this letter also, which he published in *Philosophical Transactions* in 1717). Local historians confirm that Wigan had an active metallurgical industry and also made clocks in this period, though the destruction of many of the town's records in the ensuing Civil Wars makes it hard for modern scholars to check details. Although the frames of the astronomical quadrants and sextants of this time were often of stout wood, with metal scales, pivots and perhaps bracings, it had already become common to make important pieces entirely in metal to ensure rigidity, and hence, accuracy. When visiting Towneley Hall in 1672, for example, John Flamsteed recorded being shown the 'Iron Carkasse' of the late William Gascoigne's sextant,[35] which might have been the five-foot-radius instrument that his father's over-worked blacksmith had been constructing, while in

The stained glass window of St Michael's Church, Much Hoole, depicting Horrocks observing the transit of Venus across the sun, an event he had predicted (image taken from A. Chapman, *Jeremiah Horrocks and Much Hoole*)

the Museum of the History of Science, Oxford, are four iron-framed instruments by the London master instrument maker Elias Allen – one of them dated 1638 – that are very similar in type to those being discussed by Crabtree and Gascoigne.[36]

However, even if the North Country astronomers employed country joiners, blacksmiths and brass founders to make the frames of their quadrants and sextants, they almost certainly would themselves have graduated their instruments with degree and minute of arc divisions. Anybody with the skill to make an accurate astronomical observation would have mastered this technique as a matter of course and Flamsteed recorded so graduating his own seven-foot sextant at Greenwich in 1677.[37]

It was in 1639, however, that Jeremiah Horrocks made the discovery which would immortalise both himself and Crabtree. While Kepler, in his own study of planetary dynamics, had correctly predicted that the planet Venus would pass across the sun's disc in 1631, no one had observed the event because it happened after sunset for European observers. Kepler had said that another Venus transit would not occur until 1761 but in October 1639, Horrocks, now working in some ecclesiastical capacity (though still too young to be a deacon or a priest) in the village of Much Hoole, near Preston, Lancashire, calculated that on 24 November 1639 Venus would once again transit the sun. He told Crabtree and other astronomical friends, but only he and Crabtree saw the event on that cloudy late November day. For Horrocks had found that the celebrated Kepler had made an error in the orbital theory of Venus; he corrected that error, and, along with Crabtree, became the first person on the face of the earth to witness a Venus transit. Horrocks then wrote a short Latin treatise describing his and Crabtree's work in which he reminded his readers of the importance of not simply following others but of relying on one's instruments and observations if one wanted to make original discoveries and to advance scientific knowledge. He also used his *Venus in sole visa* ('Venus in transit across the sun') as a manifesto for Copernicanism.

By the time of the composition of the *Venus* treatise in 1640 and from the great mass of correspondence which had passed between them, one can see how far Horrocks and Crabtree had moved away from being students and disciples of Kepler and Galileo to being independent, original researchers. Well in advance of any of their English contemporaries, they had come to grasp the *practical* significance of how to do original astronomy. They had come, in the wake of

St Michael's Church, Much Hoole

experience, to mistrust the standard tables and ways of working, and to realise that creative science did *not* lie in the occasional checking, confirming or polite correcting of what their illustrious forebears had already laid down, but in throwing away the rule book. Astronomy, indeed, could only be advanced by the exhaustive and meticulous observation and re-observation of the heavens on a nightly basis, conducted with instruments whose accuracy parameters had been established beforehand, and which could then be improved upon and made to cross-check each other. Sextants, cross-staves, quadrants, telescopes and micrometers would be used to construct a matrix of observational data to quantify the physical parameters of the terrestrial (from the sun), lunar and Venus orbits, the mutual apparent acceleration and slowing down of Jupiter and Saturn, and the point-like character of the stars. And from three observations of the position of Venus on the sun's disk, made during the 1639 transit, Horrocks had (by a line of physical reasoning which we now know to have been ingenious yet incorrect) come up with a value for the solar parallax –14 arc seconds – that was a vast improvement on those of Kepler and Tycho.[38] The accepted modern value for the solar parallax is 8.80 arc seconds, giving a mean solar distance of 93,000,000 miles.

But Horrocks in particular was much more than a brilliant and meticulous observer; he saw his observational data as inextricably bound up with bigger theoretical and cosmological issues. What were the forces that made the moon and planets move in orbits of such complex yet perfect ellipticity? How do these celestial changes produce the tides? And how big is the solar system and stellar universe? It is hardly surprising that Horrocks's work influenced the development of gravitation theory especially after the surviving letters and astronomical essays were published in *Opera posthuma* in 1672, and that Horrocks receives respectful citations by Newton in *Principia* in 1687. For at heart, Horrocks was a geometer who was fascinated by the exquisite perfection of astronomy.

Horrocks and Crabtree died of natural causes in 1641 and 1644 respectively while Gascoigne fell at Marston Moor in 1644. After their deaths Richard and Christopher Towneley began to collect what survived of Horrocks's, Crabtree's and Gascoigne's books, manuscripts and instruments. But the three men's work had also inspired others in the vicinity. Jeremy Shakerley, a Lancashire client of the Towneleys, was deeply influenced by 'Our worthy Country-man, Master Jeremy Horrox' and became the first writer to mention Horrocks's name in print in his books published between 1649 and 1653.[39] Then there was the mathematical lawyer Nathan Pighells and the young Sir Jonas Moore, also of the Towneley connection. It was Royalist Moore, indeed, who would take the work of this Lancashire circle to London and, after 1660, into the ranks of the newly established Royal Society, which would in 1672 publish Crabtree's and Horrocks's surviving correspondence as *Opera posthuma*, rightly deeming it the beginning of a critical, observational and experimental tradition of British astronomy. Interestingly, the man whom the Royal Society entrusted with this very important editorial enterprise was Horrocks's old Emmanuel College contemporary, the Revd Dr John Wallis, then Savilian Professor of Geometry at Oxford, and one of the leading mathematicians in Europe.

Horrocks's *Venus in sole visa* was first published in Dantzig, Poland, in 1662. It seems that in the early Royal Society, a manuscript copy of the *Venus* treatise came into the hands of the Dutch astronomer Christiaan Huygens when he was on a visit to London in 1661. From Huygens this manuscript was passed on to Johannes Hevelius in Dantzig and Hevelius published it at his own expense in 1662. And as Huygens and Hevelius were two of the most eminent astronomers in Europe at that time, one gets some measure of the significance of the northern work.

By the early 1670s the achievements and fame of Horrocks, Crabtree and Gascoigne were doing the rounds of learned Europe, though it was a Lancastrian who became their first published biographer. Sir Edward Sherburne, Roman Catholic scholar-squire and proprietor of the Stonyhurst estates, published in 1675 his English translation of the astronomical poem of the Roman poet Marcus Manilius. *The sphere of Marcus Manilius* contained by way of an appendix 'A Catalogue of the most Eminent Astronomers, Ancient & Modern', which included the first short integrated biographical essays upon Horrocks, Crabtree, Gascoigne, Sir Jonas Moore, Shakerley and, of course, Christopher

Towneley, amongst others of the northern connection.[40] As Sherburne was a neighbour, close contemporary and a friend of the Towneleys, he probably had information 'from the horse's mouth', as it were. When John Aubrey composed that famous biographical collection subsequently immortalised as *Brief lives* around 1689, he included a 'Life' for Gascoigne.[41] According to Aubrey, Gascoigne, the 'most Gallant gentleman and excellent Mathematician that dyed in the late warres ... was bred up by the Jesuits'. Aubrey seems to have credited his information about Gascoigne directly to Sir Jonas Moore, Gascoigne's erstwhile protégé, whom one presumes Aubrey knew personally through the Royal Society dining circuit in London though Aubrey regretted that Sir Jonas's sudden death (in the summer of 1679) had prevented him from learning more.

Yet Aubrey adds further Gascoigne stories in his *Brief life* of Sir Jonas Moore, claiming that Gascoigne had been educated 'by the Jesuits in Rome [who] gave him good information in Mathematicall knowledge'. Furthermore, Gascoigne seems to have been taught 'a new way of flying' by the Jesuits which he passed on to an Irish boy and enabled the boy to fly 'over a river in Lancashire'; though as with so many of Aubrey's stories, often picked up in the course of conversations – and in the case of Sir Jonas Moore in particular, probably written down some time after its original narration – details tended to get mixed up. It would be interesting to know if Gascoigne ever attended the English Jesuit College in Rome as did so many other young English Catholic gentlemen. That would certainly have given him a command of the Latin language well in advance of the Latin juvenilia attributed to him found in the exercise books preserved in Chetham's Library, Manchester.[42] And had he spent time in Rome somewhere around 1632, when he would have been 20 years old, Gascoigne could well have known the Jesuit astronomers of the first Vatican Observatory or even have met some of the figures involved in the Galileo affair. But this is pure speculation, based on remarks in John Aubrey's sometimes wild-sounding narrative.

As was indicated at the outset, there are only partial explanations why the great European astronomical Renaissance should have taken root in what was then the remote and relatively backward north-west of England in the 1630s. The researches and personal friendships that sprang up between Horrocks, Crabtree, Gascoigne, Towneley and others begs more historical questions than can be easily answered. Particularly fascinating were their religious beliefs, for though all of them were sincere and devout Christians, Horrocks and Crabtree were Protestant Anglicans while Gascoigne and the Towneleys were from recusant families: ancient gentry families which had simply refused to acknowledge King Henry VIII's Protestant Reformation. There were many of these recusant Catholic families in Lancashire and in some cases they paid a heavy price for their faith: crippling recusancy fines to the Protestant government, severe punishment if Roman Catholic priests were found to be visiting their houses, and exclusion from the public life of their class, for recusants could not generally serve as magistrates, regular military officers or members of parliament. Perhaps one reason why the Towneleys, Gascoignes and Sherburnes developed such conspicuous scientific and cultural interests is because these were subjects from which English Catholic intellectuals were not excluded by law.

To find recusants on such amiable terms with their intellectual Protestant neighbours forces us to revise hoary myths not only about Catholics and Protestants in the seventeenth century, but also about the wider relationship that existed between science and religion. For what all of these men saw in science, in addition to its technical fascination, was an avenue for the deepening of their Christian faith as they explored the intricate wonders of God's Creation. As the devout Protestant Johannes Kepler put it, scientific discovery was but 'thinking God's thoughts after Him'.

80 years after the death of the North Country astronomers, the Revd John Flamsteed's *magnum opus*, the *Historia coelestis britannica* ('British account of the heavens') was brought out in three thick folio volumes in 1725. He commenced this, the first great star catalogue of the Royal Observatory, Greenwich, with five pages of hitherto unpublished Gascoigne observations from 1638, and correspondence that had passed between Crabtree and Gascoigne between 1640 and 1643. For Flamsteed, the first Astronomer Royal, these North Country astronomers had laid the foundation stone upon which subsequent British astronomical research would be built.

Coda: The post-1639 transits of Venus

Only five other Venus transits occurred after 1639, falling in 1761, 1769, 1874, 1882 and most recently, on 8 June 2004. In honour of Jeremiah Horrocks and his friends, the International Astronomical Union held its special Transit of Venus Conference at the University of Central Lancashire in June 2004. The beginning of the transit was observed at the University's own Observatory at Alston Hall, Longridge, Preston, after which we were all taken to Much Hoole to witness the egress. In 2004 the transit was also observed in Salford and demonstrated to the public from the Observatory in Chaseley Fields, and from elsewhere in the city. Research by Ken Irving, Carl Barry, and Lilian Fletcher has also identified a house in Broughton which seems, after modifications over the centuries, to have been the home of William Crabtree.[43] And as part of the June 2004 transit celebrations, Salford City Council renamed a lane running up to this house 'Crabtree Close'. Another transit will occur in 2012, but after that we must wait until 2117.

Acknowledgement

I wish to thank Dr Robert van Gent of Utrecht University, Holland, for information regarding the passage of Horrocks's manuscript to Hevelius by Christiaan Huygens.

Notes

The present chapter is a much extended and amplified version of a chapter published as 'To the heavens in rural Lancashire: Jeremiah Horrocks and his circle and the beginning of British astronomical research', in R. Finnegan (ed.), *Participating in the knowledge society* (London, 2005).

1. J. North, *The Fontana history of astronomy and cosmology* (London and New York, 1995); M. Hoskin, *The Cambridge illustrated history of astronomy* (Cambridge, 1997); G. Galilei, *Siderius nuncius* ['The starry messenger'] (1610), in S. Drake (ed.), *The discoveries and opinions of Galileo* (New York, 1957), pp. 1–58.
2. A. Chapman, 'The astronomical work of Thomas Harriot (1560–1621)', *Quarterly Journal of the Royal Astronomical Society*, 36, 1995, pp. 97–107; J. Shirley (ed.), *Thomas Harriot: renaissance scientist* (Oxford, 1974); J. Roche, 'The Radius Astronomicus in England', *Annals of Science*, 38, 1981, pp. 1–32.
3. A. Chapman, 'William Crabtree, 1610–1644: Manchester's first mathematician', lecture given to the Manchester Statistical Society, 13 June 1995 (Manchester Statistical Society, 1995).
4. W. Derham (ed.), 'Observations of the spots that have been seen upon the sun from the years 1703–1711, with a letter of Mr Crabtree, in the year [7 August] 1640, upon the same subject', *Philosophical Transactions of the Royal Society* [hereafter *Phil. Trans.*], 27, 1711; W. Derham (ed.), 'Extracts from Mr Gascoigne's and Mr Crabtree's letters, proving Mr Gascoigne to have been the inventor of the Telescopick Sights of Mathematical Instruments, and not the French', *Phil. Trans.*, 30, 1717, pp. 603–10; S. J. Rigaud, *Correspondence of scientific men of the seventeenth century*, 1 (Oxford, 1841), p. 35; J. Flamsteed (A. Chapman [ed. and intro.]), *The 'Preface' to John Flamsteed's 'Historia Coelestis Britannica', 1725* (based on a translation by A. D. Johnson [National Maritime Museum Monograph No. 52, National Maritime Museum, Greenwich, 1982]); A. Chapman, 'Jeremiah Horrocks, the transit of Venus, and the "New Astronomy" in early-seventeenth century England', *Quarterly Journal of the Royal Astronomical Society*, 31, 1990, pp. 333–57; R. V. Taylor, *Biographia leodensis* (1865).

5. P. Aughton, *The transit of Venus: the brief, brilliant life of Jeremiah Horrocks, father of British astronomy* (London, 2004), p. 190.
6. C. Barry and L. Fletcher, 'William Crabtree of Broughton. A hunt for his home', computer manuscript, privately produced (Salford, 2004). Copy on deposit in Salford City Central Library, Peel Park, Salford.
7. 'William Crabtree of Broughton in the County of Lancaster, Yeoman': Land deed, 8 July 1619, for property in Levenshulme, Manchester. Crabtree co-signatory along with Richard Leghe, Henry Leghe, and Ralph Cartwright. City of Manchester Local History Reference Library MS.
8. 'Book of Rates for the County Pallatine of Lancaster.' Signed W. Crabtree. Undated, but probably post-1644, and not William the astronomer. Chetham's Library MS., Manchester.
9. William Crabtree of Broughton: Will, 19 July 1644. Proved Canterbury under 'Rivers', PRO PROB11, 194, 240.
10. James Crossley (ed.), *The diary and correspondence of Dr John Worthington*, Chetham Society, 13 (Manchester, 1847).
11. J. Wallis (ed.), *Opera posthuma; being Jeremiae Horrocci ... Opera posthuma* (London, 1673), p. 247 [mis-paginated as p. 347].
12. Crabtree, *Phil. Trans.*, 1717, pp. 603–10.
13. In a letter which Horrocks wrote to Crabtree on 26 Oct. 1639, informing him of the forthcoming transit of Venus on 24 Nov. 1639, he begged Crabtree to 'apprise Mr. Foster' if time permitted, thereby suggesting that while both Horrocks and Crabtree knew Foster, it was Crabtree who had the more regular communication with him: *Opera posthuma*, p. 331. Indeed, it is clear from Crabtree's 7 Aug. 1640 letter to Gascoigne that he and Samuel Foster maintained a close working relationship and correspondence, for he told Gascoigne that not only was he going to pass on his own and Horrocks's observations of a recent eclipse to him, but also 'Mr Foster's at London': *Phil. Trans.*, 1711, pp. 280–89, at p. 288.
14. A. Chapman, 'Gresham College: scientific instruments and the advancement of Useful Knowledge', *Bulletin of the Scientific Instrument Society*, 56, 1998, pp. 6–13.
15. Rigaud, *Correspondence of scientific men*.
16. C. Webster, 'Richard Towneley, 1629–1707, and the Towneley group', *Transactions of the Historic Society of Lancashire and Cheshire*, 118, 1966, pp. 51–76.
17. H. Power, *Experimental philosophy* (London, 1663), pp. 121–7.
18. R. Towneley, of Towneley Hall, Burnley, Will, 11 April 1706: Lancashire County Record Office, Preston, WCW.
19. Flamsteed, *Historia coelestis britannica*, Preface.
20. Flamsteed, *Historia coelestis britannica*, i, p. 1.
21. Chapman, 'Jeremiah Horrocks', p, 344.
22. Wallis (ed.), *Opera posthuma*, p. 331.
23. *Phil. Trans.*, 1711.

24. R. Towneley, 'A description of an instrument for dividing a foot into many thousand parts, and thereby measuring the diameters of the planets to great exactness', *Phil. Trans.*, 2, 1667, pp. 541–4; A. Chapman, *Dividing the circle: the development of critical angular measurement in astronomy, 1500–1850* (Chichester, New York, 1990; 1995), p. 40.
25. J. E. Bailey, 'Jeremiah Horrocks and William Crabtree, observers of the transit of Venus, 24 November, 1639', *The Palatine Notebook*, 2, 1882, pp. 253–66; J. E. Bailey, 'The writings of Jeremiah Horrox and William Crabtree', *Palatine Notebook*, 3, 1883, pp. 17–22; Aughton, *The transit of Venus*, p. 63.
26. Wallis (ed.), *Opera posthuma*, pp. 362–3.
27. J. Wilkins, *Discovery of a new world ... in the moon* (1640).
28. A. Chapman, 'Horrocks, Crabtree, and the 1639 transit of Venus', *Astronomy and Geophysics*, 5:45, 2004, pp. 26–31.
29. Chapman, *Dividing the circle*, pl. 14.
30. E. Axon, 'William Crabtree's plan of the Booth Hall estate', *Transactions of the Lancashire and Cheshire Antiquarian Society*, 23, 1905, pp. 30–3, with fold-out reproduction of Crabtree's beautiful estate plan, and entitled 'A TRUE PLOTT or Topographical Description of one messuage and Tenement of MR HUMFREY BOOTH'S lyinge in Blakeley in the Countie of Lancaster', and 'Performed by WILLIAM CRABTRIE Anno Domini 1637'.
31. Flamsteed, *Historia coelestis britannica*, i, p. 4.
32. F. A. Bailey and T. C. Barber, 'The seventeenth-century origins of watchmaking in west Lancashire', in J. R. Harris (ed.), *Liverpool and Merseyside* (London, 1969), pp. 1–15.
33. *Phil. Trans.*, 1717, pp. 603–10, at p. 605.
34. Towneley, 'A description'.
35. Chapman, *Dividing the circle*, pp. 36–7.
36. A. Chapman, 'The design and accuracy of some observatory instruments of the seventeenth century', *Annals of Science*, 40, 1983, pp. 457–71; reprinted in A. Chapman, *Astronomical instruments and their users: Tycho Brahe to William Lassell* (Aldershot, 1996).
37. See Chapman, *Dividing the circle*.
38. See Chapman, 'Jeremiah Horrocks', p. 355, n. 76.
39. A. Chapman, 'Jeremy Shakerley, 1626–1655(?)', *Transactions of the Historic Society of Lancashire and Cheshire*, 135, 1986, pp. 1–14.
40. E. Sherburne, *The sphere of Marcus Manilius* (London, 1675).
41. J. Aubrey (A. Clark [ed.]), *'Brief lives' chiefly of contemporaries, set down by John Aubrey between 1669 & 1696* (Oxford, 1896).
42. W. Gascoigne, MS notebook. Undated. Containing early exercises, popular aphorisms, and astronomical translations and computations from Philip Lansberg. Chetham's Library, Manchester, MS A 3110.
43. Barry and Fletcher, MS., 2004.

Richard Roberts (1789–1864), pioneer of production engineering in Manchester

Richard Hills

For many of the years following the ending of the Napoleonic Wars, Manchester was a boom town. There were set-backs, such as at the time of the Peterloo riots in 1819 and again after 1825, but generally the cotton textile industry continued to expand with production costs lowered through improved machinery. Then the opening of the Liverpool and Manchester Railway in September 1830 marked the start of a new phase in the Industrial Revolution. The spread of inter-city railways created fresh markets not only for railway equipment such as steam locomotives, rolling stock and permanent way but also the where-with-all to construct and maintain everything. Richard Roberts was one of the leading inventors of his day who not only contributed improvements to textile machinery and locomotives but also introduced new types of machine tools and inaugurated new production methods for making many replicas of an artefact.

Richard Roberts first came to Manchester in 1814, seeking work to escape the militia officers.[1] His last job had been at Horseley Iron Works, Tipton, as a leading hand in the pattern shop. He had been taught woodwork and lathe-turning in his home town of Llanymynech by the local priest, the Revd Griffith Howell, who had caught the lad carving a walking-stick instead of attending to his lessons during school. Roberts received only a rudimentary education at classes held in the church belfrey where he was considered a dull pupil, but luckily Howell realised Roberts's practical aptitudes and encouraged him.

Roberts was born on 22 April 1789 in the New Bridge Tollhouse on the borders of England and Wales where his father William combined shoe-making with taking tolls. While Roberts later considered himself a Welshman, this in fact depended on the room in which he was born as the country boundary ran through the middle of the house. He was the second of four sons and there were three younger sisters. He made a spinning wheel for his mother and also repaired a lathe for someone, so he became quite proficient and later claimed that he excelled any other person in the art of wood turning.

He started work with a boatman on the newly-opened Ellesmere

Canal before being employed in the Llanymynech limestone quarries where his fellow workmen bought him a set of tools. A local road surveyor, Robert Baugh, gave him lessons in drawing, which were invaluable to him later. At the age of 20, Roberts moved to the Bradley Iron Works, Bilston, as a patternmaker, and then on to Horseley. To escape the militia he said farewell to his father at Llanymynech and headed for Liverpool. Finding no work there, he decided to walk on to Manchester where he arrived at dusk, weary and dirty in consequence of the road being in a state of mud and ruts.

Richard Roberts, 1789–1864

There are two stories about his first experience in Manchester. One is that he went to an apple stall to buy a pennyworth of apples but, in reality, to ask about employment. The stall-holder knew of a wood-turner who wanted a hand. Luckily the turner was in and offered Roberts a job for the following morning. The other story is that Roberts went to the White Lion inn with only enough money for a glass of beer. He was introduced to an old turner who wanted someone to turn the flywheel of his lathe which Roberts undertook to do, and the landlord of the White Lion offered him lodgings for a few nights on trust. On the following Monday, the workman, who was carving bedposts on the lathe, failed to turn up and, because the turner's eyesight was too poor to do the job, Roberts did it. After a short while there, Roberts moved on to another job in Salford at lathe- and tool-making but, hearing that the militia officers were still searching for him, he decided to move to London.

Roberts walked south with Francis Lewis and a person called Murgatroyd. There was a mutual understanding that whoever obtained employment first would divide his earnings with the others. Roberts was for a short time at Holtzapffel's, the famous lathe-maker, before moving on to Henry Maudslay, the leading mechanical engineer at this time. Maudslay had become famous through building the range of tools for making ships' pulley blocks designed by Marc Brunel. Maudslay had cut accurate screw threads and introduced the slide rest on the lathe which enabled the cutting tools to be held more accurately and firmly than by hand. He had made a very accurate micrometer which he nicknamed 'The Lord Chancellor' because on it could be judged the accuracy of any other jobs produced in his works. Roberts learnt a great deal from Maudslay about different types of machine tools, the need for accuracy in mechanical engineering, as well as the standards in quality at which Maudslay aimed both in design and manufacture. It is not clear why Roberts decided to leave Maudslay. Perhaps he could not exploit his own inventive genius. But once the Napoleonic Wars ended in 1815 and the threat of being conscripted into the militia receded, Roberts left London.

Roberts's early machine tools

Roberts had returned to Manchester by 1816 and set up in business as a turner of plain and eccentric work at 15 Deansgate. The lathe was upstairs in a bedroom, driven from a big wheel in the basement by his wife. When, and to whom, he was married remains unknown. Likewise, they appear to have had no children and her death is unrecorded. Holtzapffel had representatives across the country to sell his lathes and possibly Roberts was one. When Charles Holtzapffel wrote his book *Turning and mechanical manipulation* in the 1840s, it is clear that he was well aware of Roberts's improvements to machine tools and so must have known about Roberts's inventions. Holtzapffel's lathes were designed for amateur gentlemen and Roberts may have found little demand for them in Manchester. But in the 1819 *Manchester Directory*, Roberts is described as 'Lathe and Tool Maker', with his workshop at the New Market Buildings, Pool Fold, and his private house at 5 Water Street. What is clear is that, by 1819, he had already established himself as an engineer, inventing a range of much heavier machine tools.

Although there were many other engineering firms in Manchester and Salford at that time, such as Bateman and Sherratt, T. C. Hewes or Peel, Williams, Peel which were capable of building steam engines, textile machines and waterwheels, their machine tools were still very primitive. William Fairbairn later claimed that in 1817 'even Manchester did not boast of many lathes or tools, except small ones in the machine shops; ours was of considerable dimensions, and capable of turning shafts of from 3 to 6 inches diameter',[2] implying that his lathe was superior to the others. Roberts's contribution lay in making machine tools that had greater accuracy, in which the cutting tools were no longer held manually but were mounted on toolposts moved by screw threads along flat slides.

The first of Roberts's new machine tools was a gear-cutting machine which suggests that he had identified a market for more accurately cut gearwheels in the textile industry. Machines to prepare cotton for spinning, such as carding engines and roving frames, all needed gearwheels, as did the actual spinning machines like Richard Arkwright's waterframe and Samuel Crompton's mule. Well formed gearwheels would have given greater steadiness to the running of these machines and hence greater uniformity to the yarn.

Roberts said later that he had a set of change-wheels made in 1816 by a clockmaker which were so much out of pitch that he investigated the error. He discovered that the measuring device, a sector, was inaccurate, so he set about contriving his own. He sold these in sets for gearwheels of different diameters and different pitches of teeth, and must have had means of engraving these with the necessary scales.

The change-wheels may have been destined for Roberts's gear-cutting machine. Watch- and clockmakers had long had tools for cutting the teeth in gear blanks. The blank was mounted horizontally on top of a vertical arbour with a dial plate below which could be rotated one notch at a time, corresponding with the number of teeth needed. A shaped rotary cutter was lowered to cut the slot between each tooth. We do not know the design of his first gear-cutting machines, but in 1821 Roberts advertised a new and improved version; one has been preserved in the collections of the London Science Museum, ascribed to Collier until recently. Roberts up-ended the clockmakers' machine so the gear blank is mounted vertically on the arbour. At the back is a large indexing wheel turned by a wormgear. Changeable intermediate gearing allows the operative to give one turn of the handle to advance the blank one tooth. The rotary cutter is advanced under the blank to cut the slot between the teeth. The table can be set to give an angled cut through one adjustment, or through another can be tilted for cutting bevel gears. Scales engraved on the machine show the correct angles. It is an impressive machine, even more so if it originates from 1816 because its size is so much greater than contemporary ones for making clocks.

In the following year, Roberts invented his planing machine. Priority for this has been claimed by others. He was certainly

Roberts's advertisement in the first issue of the *Manchester Guardian*, 5 May 1821

RICHARD ROBERTS,
Lathe, Screw, Screw-Engine, Screw Stock, &c., &c., Manufacturer,

RESPECTFULLY informs COTTON-SPINNERS, IRON-FOUNDERS, MACHINE-MAKERS, and MECHANICS in general, that he has CUTTING-ENGINES at work on his NEW and IMPROVED principle, which are so constructed as to be capable of producing ANY number of Teeth required : they will cut BEVIL, SPUR, or WORM Geer, of any *size* and *pitch* not exceeding 30 inches diameter, in WOOD, BRASS, CAST-IRON, WROUGHT-IRON, or STEEL, and the TEETH will NOT REQUIRE FILEING UP; DIVISION-PLATES, QUADRANTS, &c., accurately divided, or additional Numbers put on *Old Plates.*

N.B. R. R. cuts, on his *Improved Screw Engine,* SCREWS of ALL Sorts, Pitches, or sizes, with the greatest accuracy.

Manufactory, New Market Buildings, Pool Fold; House, 5, Water-street, Manchester.

ignorant of these other machines and later *The Engineer* supported him as being the first inventor. Roberts had been making copying presses for James Watt's method of duplicating letters by pressing a sheet of damp thin paper onto the original written with a mucilaginous ink that transfused into the second sheet under pressure. The person who chipped and filed smooth the head and bottom faces of the cast iron presses charged exorbitant sums which he spent on drink and so was irregular in his work; Roberts decided to replace him with a machine tool. His original planing machine is also preserved in the Science Museum. The table, which moved under the cutting tool, is 52 ins long by 11 ins wide so would have been capable of planing only a small copying press but would have been more suitable for lathe beds.

Roberts sold a planing machine to Maudslay in 1826 which marks the beginning of the introduction of these tools to engineering workshops. He turned them into extremely versatile and useful tools. Not only could they make flat horizontal surfaces but vertical and angled ones as well. By using shaped tools, surfaces on the bottoms of parts could be machined and also mouldings. Spiral and curved surfaces could be machined by mounting headstocks or secondary tables on the main table and rotating the workpiece.

Another of Roberts's machine tools preserved in the Science Museum is his large lathe dated to 1817. It contains many novel features such as back gearing to increase its range of speeds, another of Roberts's inventions. The cutting tool is securely mounted on a saddle hung on the front of the bed so it can traverse and cut right up to the ends of the work piece and does not limit the diameter of the part to be machined. The saddle could be moved either manually or by the mandrel in the headstock turning a dial plate with rings of teeth to give seven different speeds. A dog clutch engaged one of a pair of bevel wheels to move the saddle in the appropriate direction. The saddle hit an adjustable stop on a horizontal rod which disengaged the clutch, making the lathe partially self-acting, but the rate of traverse is too slow for screw cutting. This must be considered as a very early true industrial lathe for it was capable of quite heavy work.

Another early machine tool was his lathe for cutting long screw threads. Interchangeable master threads of various pitches could be mounted on it to move the saddle holding the cutting tool. A special holder enabled the tool to be angled according to the helix of the thread. These features, together with changeable gearwheels, enabled a wide range of threads to be machined on bar-stock. So useful was this tool that Charles Beyer purchased it together with the large lathe for use in his new works of Beyer, Peacock & Co. in 1854. The screw-cutting lathe was employed until well after 1900 for producing the threads on locomotive reversing mechanisms.

These and other machine tools laid the basis for Roberts's development of production engineering techniques and impressed Mancunians for, in 1817 or 1818, he was asked by the Commissioners of Police to invent a gas meter which could measure the amount consumed by their customers. Roberts quickly produced two within a week on which he solved the problem of preventing gas escaping, by use of a water seal. Samuel Clegg in London had failed to do this and quickly copied Roberts's method because Roberts could not afford to take out a patent. We do not know who made these meters for the Manchester Police Commissioners but they show that Roberts could design intricate mechanisms as well as large machines.

The workshop in Pool Fold was in a poor situation in the centre of Manchester and the major part of it was on an upper floor reached by a Jacob's ladder. Presumably Roberts, who must have returned to Manchester with virtually no capital, financed developments out of any profits. In 1818 he made a breech-loading rifled brass cannon on which a Mr Bradbury experimented with lead-coated spherical missiles. Rifling and breech-loading became common only during the Crimea War. In August 1821, Maudslay's partner, Joshua Field, called on Roberts during a visit to Manchester. Roberts took him to see where he had escaped from the Peterloo Massacre. In Roberts's workshop, Field noted a couple of forges downstairs, and upstairs, three smaller lathes as well as the 1817 one, drilling- and screw-cutting machines and a gear cutter, all of which were driven from lineshafting turned by three men. All told, Roberts was employing 12 or 14 men, so he had considerably expanded his scale of operations.

Textile machines

Field commented on Roberts's improving a reed-making machine, patented in Britain by an American, Jeptha Avery Wilkinson, in 1817. It produced the reeds for looms, devices like a comb to keep the warp threads spaced apart correctly and also to beat up the weft. Somehow the rights had been obtained by four people living in Manchester, Thomas Sharp and Thomas Jones Wilkinson and probably Robert Chapman Sharp and James Hill. T. Sharp, Hill and T. J. Wilkinson were involved in the iron trade which may be how Roberts knew them. T. J. Wilkinson told Matthew Robinson Boulton in 1823 that he thought he, Boulton, would be 'gratified with the concern generally & with the mechanical genius of Mr. Roberts who has the management of that department'.[3] By this time, not only had Roberts improved preparation of the iron strip for the dents in the reed but had also improved J. A. Wilkinson's machine so that reeds were made more accurately and quickly than by hand. Unfortunately no information has survived about the design. Small intricate mechanisms became one of Roberts's specialities. In this case, he mechanised a series of operations with some controlling device which would start and stop each particular operation. In this way, his reed-making machine foreshadowed his success with the power loom and the self-acting spinning mule and is an early example of production engineering techniques. A firm, Sharp, Hill & Co., was established to manufacture reeds. The partners seem not to have sold these machines to others in Britain but derived considerable profits from manufacturing reeds on a site in Faulkner Street, which was eventually merged into the Globe Works of Sharp, Roberts & Co.

Power looms

These reed-making machines may have turned Roberts's attention to weaving and looms so that he patented improvements to power looms in November 1822.[4] This was probably financed by James Hill who formed a partnership with Roberts as Roberts, Hill & Co. The patent covered ways of weaving simple patterns as well as superior mechanisms to unwind the warp from the beam at the back and to wind the cloth onto the front one. Roberts's patent was taken out soon after Fairbairn had started to introduce light wrought iron shafting in mills which replaced cumbersome cast iron and enabled speeds to rise and more power to be transmitted. It is probable that Roberts's better designed and more acurately manufactured looms could withstand

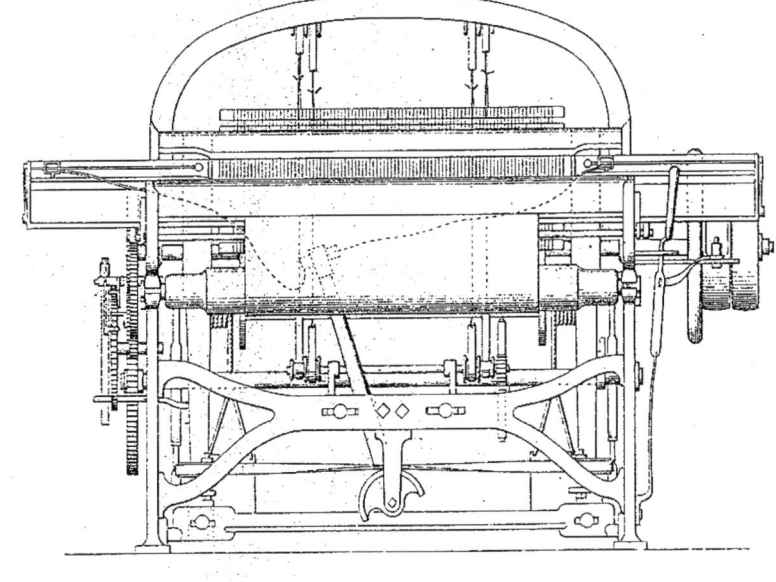

Roberts modified his power loom for weaving plain calico cloth by substituting a simpler system for working the healds and placing the picking stick at the bottom

these higher speeds. The power loom was another instance where a sequence of operations had to follow each other, with the appropriate shed being opened in the warp before the shuttle was driven across. This was followed by the reed beating up the weft, the shed being changed and the cloth being advanced a fraction. In this case, the partners decided to produce the looms themselves for sale and not manufacture cloth. By 1825, 80 were being manufactured a week, or 4,000 a year. Around 1828, Chorlton Mills had a loom shop by the Medlock with 600 looms and larger sheds were established elsewhere so that the price of coarse cotton cloth fell by 70 to 80 per cent. This rate of loom production caused Roberts to look at ways of quicker, better manufacture with specialised machine tools and led to his first attempts at batch production and production engineering. He would have devised special templates and jigs for making standard parts on a large scale which could be held in stock for quick assembly when required.

Roberts's standing in Manchester

Success with his reed-making machines and power looms must have enhanced Roberts's reputation in Manchester. In 1823, he was elected a member of the Manchester Literary and Philosophical Society where he would have met many leading Mancunians. He was elected an Honorary Member in 1861. The Society meetings would have given him the chance to discuss ideas with other engineers like Fairbairn and Peter Ewart, and it is during this period that we know about experiments by Roberts which would then have been called scientific. Roberts was constructing a foundry for the Globe Works in 1823. A little later, to his surprise, he found a valve he was fitting on the flange of an air pipe was not blown off but was retained a short distance away from the end. After a series of experiments with Thomas Hopkins to determine the force and how this was caused, Hopkins presented a paper to the Society on the 'Disc Phenomenon' in 1828. The pressure of the jet of air was not uniform across the face of the disc. It was shown that an area of low pressure existed near the perimeter.

The need for better communications between Liverpool and Manchester had long been recognised and this was the period when railways were being actively canvassed. One question raised was whether friction of the rolling stock on railways would increase rapidly with rising speed and so lead to a great increase in power needed to haul it, as happened with barges on waterways. Articles in the *Manchester Guardian* over Christmas 1824 suggested this was not the case. A puzzled Roberts built a test model. A small waggon was attached to a spring balance and placed on a three foot-diameter drum. The drum could be rotated at known speeds when it was found that the reading of the spring balance did not change with increasing speed but only if the load in the waggon was altered. Roberts's results were published in the *Manchester Guardian* on 12 February 1825 and may have helped to convince people that a railway would be a viable proposition.

During 1824, Roberts had been one of the leading members of the committee which launched the Mechanics' Institution, later the University of Manchester Institute of Science and Technology. He must have regretted that he had received only an elementary education and wished that others should have the opportunity to improve theirs. At a meeting held in the Bridgewater Arms on 7 April 1824, Roberts seconded the resolution proposing the foundation. He became one of the first directors and remained a member of various committees for many years, donating experimental apparatus which included his model waggon for testing friction.

The Globe Works

It must have become obvious quickly that the Pool Fold workshop would not have the capacity to build Roberts's power looms. In the autumn of 1822, Roberts, Hill & Co. were advertising for turners, filers and apprentices at their premises on Faulkner Street. The Globe Works was bounded on its longer sides by a branch of the Rochdale Canal and Dickinson Street, and at the northern end by Faulkner Street. This part of the works now lies under Portland Street. By July 1825, a foundry occupied the southern end of the site with forges alongside Dickinson Street. The northern part had a dining room for the men while the offices faced Faulkner Street. Three multi-storey buildings fronted onto the canal and were built with fire-proof structures consisting of cast iron columns and beams with brick arched floors.

Then disaster struck on the afternoon of Sunday 17 July 1825. A lad, walking along a near-by street, saw smoke rising and went to investigate. A couple of ladies sitting outside their house thought there was nothing suspicious because they had recently seen a clerk coming out from one of the offices. However they went along Dickinson Street, looked under the main gate and saw stacks of timber ablaze. The clerk was summoned who called the fire brigade but, in spite of a fire engine being housed only 80 yards away, horses could not be found so it was twenty minutes before the engine arrived.

The fire quickly spread to an adjoining store room with patterns and parts for power looms. Then the central fire-proof multi-storey building, which contained the joiners' shop and more stacks of timber, caught fire as well as the range of forges. Trap doors for hoists above

Banck's map of Manchester, 1831, showing the situation of the Globe Works at A. The later Atlas Works was at B

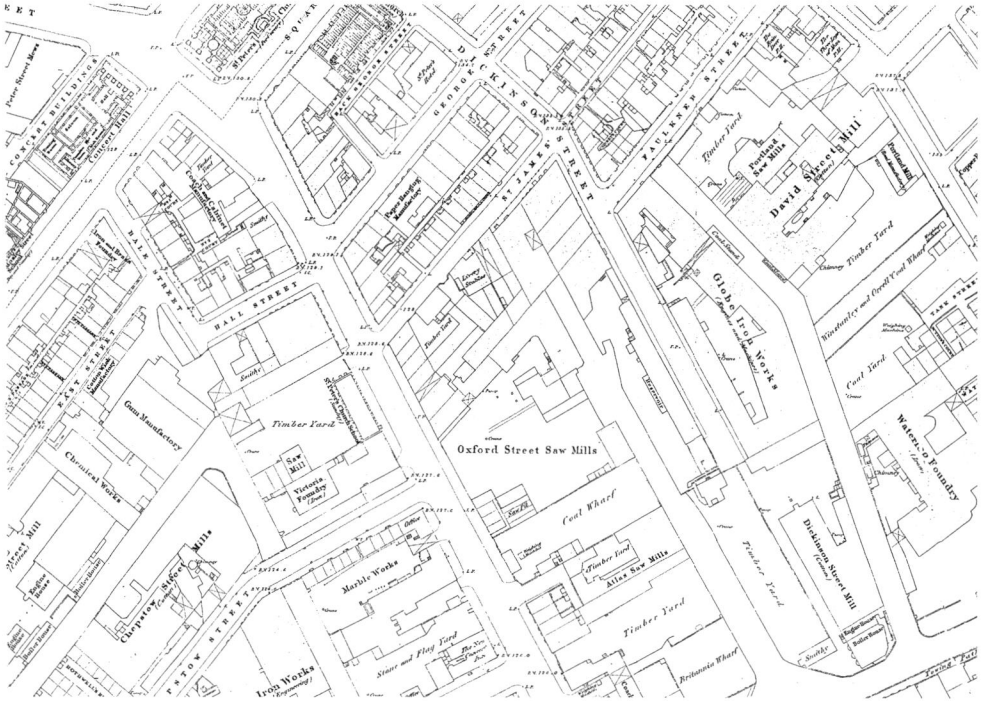

Plan of the Globe Works in 1849

the joiners' shop enabled the fire to reach the upper storeys with machine shops and the roof. The intensity of the fire caused this middle building to collapse. All the roofs in this range by the canal were wooden and caught fire. That of the third building was destroyed but, since this building had not been completed, it suffered no further damage. Luckily only the roof of the first fire-proof building was destroyed, which slightly damaged the reed-making machines below it. The fire-proof structure saved Roberts's self-acting spinning mule on a lower floor which had been completed the previous day. Valuable machine tools below that were unaffected.

Arson was suspected but never proved. Men had to be laid off and, in spite of being insured, the fire cost the company £10,154. The account in the *Manchester Guardian* stressed that the collapse of the middle building was caused by the exceptional circumstance of so much wood being stored in it and that normally such buildings would not have failed in this way. The damaged buildings were rebuilt and extended so that production was resumed by the autumn.

The spinning mule

Earlier that year, in March 1825, Roberts patented his most famous invention, the self-acting spinning mule.[5] In 1790, William Kelly at New Lanark had applied power to the spinning sequence of Crompton's mule invented in 1779. In Kelly's mule, the drafting rollers rotated, paying out the cotton fibres, the carriage moved out and the spindles turned to twist the fibres into yarn without the spinner touching the machine. But Kelly and many others after him failed to power backing-off the yarn from the spindle tips and winding it onto the spindles with the faller-wire in the form of a conical cop as the carriage moved in again. However these unsuccessful attempts particularly by William Eaton in 1818 may have helped to point Roberts in the right direction.

The mule was the most important spinning machine at this period because it could spin high quality yarn, particularly the finer types which could not be done on the waterframe. Winding-on was a skilled operation because the finished cop would be placed directly into the weaver's shuttle where it had to unwind as the yarn was pulled off evenly and without jamming or breaking. The spinners who controlled this operation could bring the whole mill to a halt if they went on strike. This was the situation which the Ashton and Stalybridge mill owners faced when trade was booming in 1824. Having heard of Roberts's inventive genius, they approached T. Sharp to see if he could persuade Roberts to invent a self-acting mule which would dispense with skilled operatives. Roberts retorted that he knew nothing about cotton spinning. Perhaps he had some sympathy for the spinners and did not wish to see the mill owners put in a commanding position. On their third visit, he relented and, after a period of four months, he brought out his first ideas.

This patent in 1825 covered three important mechanisms. First, he arranged for the sequence of spinning, backing-off and winding-on to be brought into operation in their correct sequence by the rotation of a camshaft in four stages. Also the movement of the carriage itself operated some of the mechanisms. Second, the shape and build-up of the cop was determined by a shaper rail with two adjustable inclines on it. Lastly, the tension of the yarn during winding-on was determined by the relative positions of faller- and counter-faller wires. Although the *Manchester Advertiser* carried a glowing account of the operation of this self-acting mule in November 1825, in the event, the action of the faller-wires proved to be unsatisfactory and few mules were made.

However, Roberts was not one to suffer defeat and returned to the challenge until he patented an almost perfect solution in 1830.[6] This was his famous quadrant which remained a standard feature on most mules until the last ones built for cotton spinning in 1927. In 1891, Richard Marsden wrote,

> The first self-acting mule was one of the greatest triumphs of mechanical genius that has ever been achieved, and as a display of power of the inventive faculty in man's nature surpassed anything accomplished up to that time. This statement hardly requires even that limited qualification, as though great advances have since been made in many branches of mechanical industry, nothing yet surpasses the spinning mule in the number and variability of its actions, the admirable concert of its parts, or the excellent results achieved by it ... The new machine was a perfect spinner; that is it accomplished every part of the process without manual help.[7]

A spinner on a semi-powered mule would have to push the carriage in and wind-on the spun yarn around 2,400 times to form a full cop. In contrast, the person in charge of one of Roberts's self-acting mules merely had to make adjustments to the setting of the quadrant nut to wind-on evenly. Because his mules were power-driven, more spindles could be added. Earlier hand mules had less than 200 spindles while the last self-acting mules had over 1000 so productivity per operative increased. Bigger mills were built to house these large mules. Cops spun on Roberts's mule were more regular and firmer so that there were fewer breakages during weaving, again raising productivity.

Yet few mill owners were willing to try them at first, possibly because they were very complex machines and must have needed

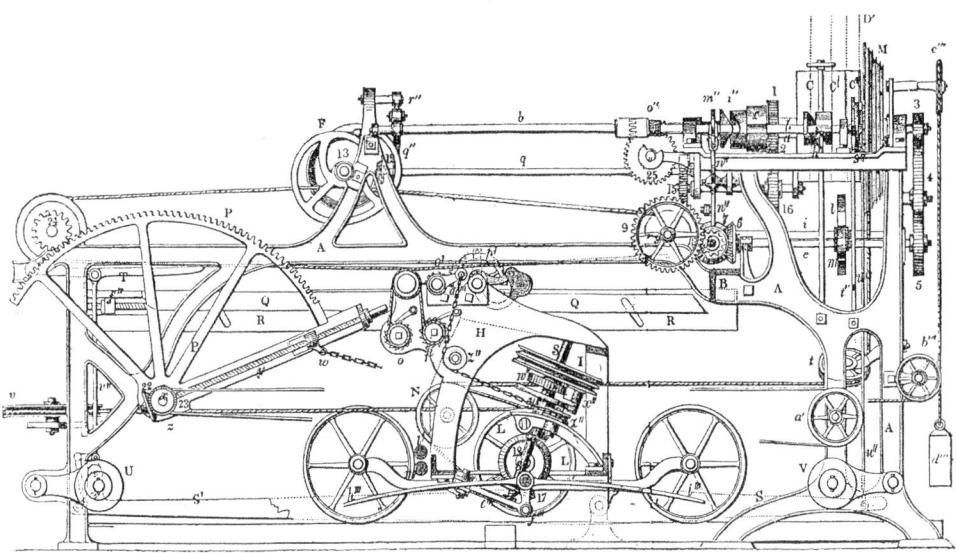

Vertical section through the headstock of a mule, based on Roberts's 1830 patent. The headstock and camshaft are top right. The carriage is in the centre with the quadrant P on the left

skilled fitters to keep them properly adjusted. But by December 1837, there were around 500,000 spindles on self-acting mules in some 100 mills. However, for Sharp, Roberts, the mule had not proved profitable. £12,000 had been spent on the 1830 patent alone but only £7,000 had been recovered in profits. Roberts's genius had so far brought little pecuniary reward. An appeal was made to the Privy Council in 1839 for the term of the 1825 patent to be extended which was granted for a further seven years.

There were changes to the partnerships in the interval between Roberts's two mule patents. Hill retired on 31 May 1826, when the firms of Roberts, Hill & Co., and Sharp, Hill & Co., were dissolved by mutual consent and replaced by that of Sharp, Roberts & Co. The remaining partners were Thomas and R. C. Sharp, T. J. Wilkinson and Roberts. They were joined by T. Sharp's younger brother John. T. J. Wilkinson and R. C. Sharp retired at the end of December 1836. At some later date, Thomas's son, Thomas Beatt Sharp, joined the firm. Thomas died on 21 April 1841, but the firm was not wound up until 24 June 1843. Roberts retained the Globe Works and Sharp Brothers the Atlas Works.

New production methods

The demand for power looms must have necessitated new production methods with specialised machine tools. All the pulleys on the line shafting and those on looms together with the gear wheels had to be secured to their shafts with keys driven into slots in their hubs.

In 1824, Roberts invented the key-grooving engine to replace a man laboriously chiselling and filing this slot. A crank and connecting rod drove a chisel up and down. The wheel to be slotted was placed on a table underneath and advanced so that the chisel gradually deepened the slot. This machine tool was improved into the slotting machine on which the table could be moved sideways as well as front to back and the table rotated. This enabled straight or curved slots to be cut and various other shapes pared out. Another tool originating from these was the shaping machine in which the cutting tool moved horizontally over the workpiece rather like filing. Here again, the table on which the work was mounted could be moved horizontally or vertically and the work rotated, so enabling the machining of complex shapes. It was noted that,

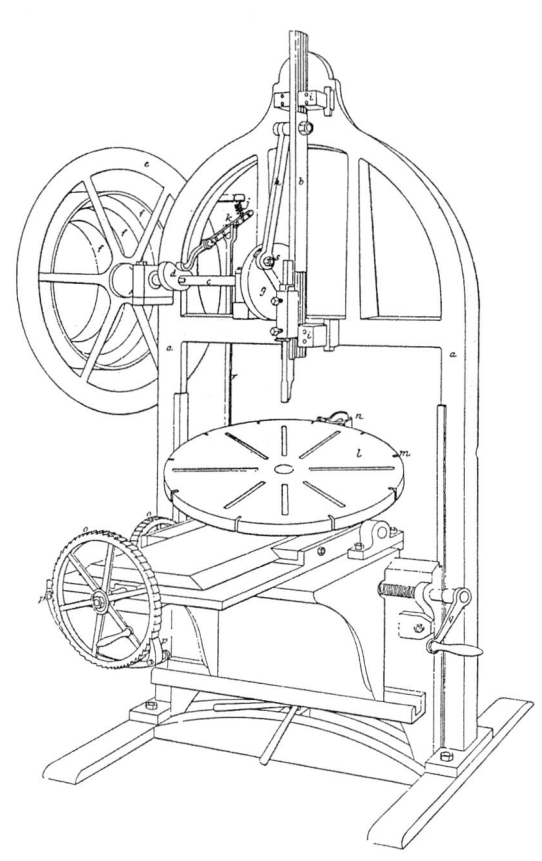

Roberts's slotting machine

> It is constructive machinery of this class which now executes a very large proportion of what used to be committed to manual efforts of the mechanic – and thus the shaping machine has virtually driven the chisel, the metal saw, and the file from the workshop.[8]

Even though each part still had to be machined individually, it was necessary that they should all be made as near identical as possible to enable a machine to be assembled with the minimum of fitting and filing. In 1825, Roberts anticipated similar large demand for his self-acting spinning mule as for his looms which in fact did not materialise at the time. So that standard parts could be made, he introduced a range of templates and gauges. He also made male and female standard gauges, or plug and collar gauges, for checking sizes of holes and bearings. A textile mill had to be equipped not only with many examples of the same machine but with many identical parts on those machines such as drafting rollers or spindles. Therefore the textile industry presented the possibility for the development of production engineering with parts being made in identical batches if not in mass production systems. *The Engineer* commented that '[t]he system of templates and gauges formed a great addition to the resources of mechanical engineering'.⁹

Steam engines for industry and transport

The opening of the Liverpool and Manchester Railway in 1830 caused Roberts to turn his attention to steam engines. In 1832, he patented an advanced system of controlling the cut-off on inlet valves for stationary beam engines.¹⁰ This patent also contained ideas for road vehicles such as differentials, steam brakes and stronger iron wheels. Roberts proceeded to build a large steam road carriage seating 35 people, perhaps his answer to the steam railway for road transport. The first trial along Oxford Road on Monday 9 December 1833 carrying 50 people showed that the boiler failed to generate enough steam but a speed of 12 mph was attained. After modifications, a further trial was held in the evening of Thursday 27 March 1834 when the boiler feed pumps were discovered not to be working. These were still not working on another run with 40 passengers on Friday 4 April 1834. They had gone a mile and a half beyond All Saints church when the fire had to be dropped through shortage of water in the boiler. It was refilled from a pond and the carriage began to return to Manchester. At the end of Rusholme Road the boiler exploded, scalding four men who had been hanging on the back of the carriage in spite of several warnings. Glass was broken in near-by windows and a shop set on fire. The carriage had to be hauled back to the Globe Works by horses, never to run again.

Roberts was not successful with his early designs of railway locomotives either. The board of the Liverpool and Manchester Railway agreed to try the first of Sharp, Roberts's locomotives on 20 May

In 1856, Roberts designed a horse-drawn omnibus for London

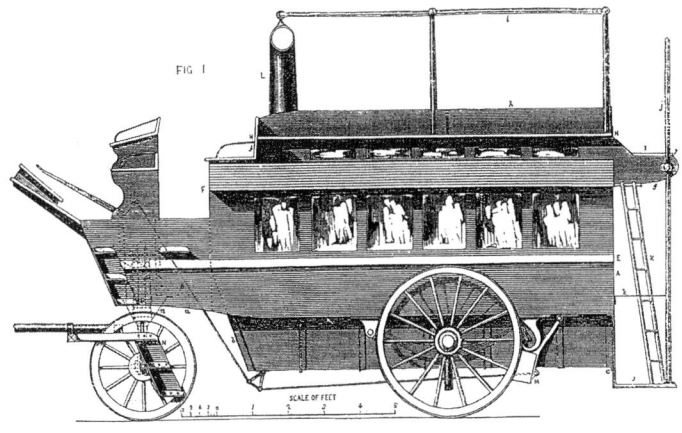
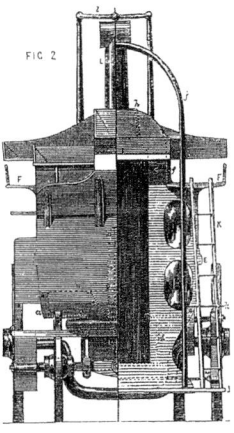

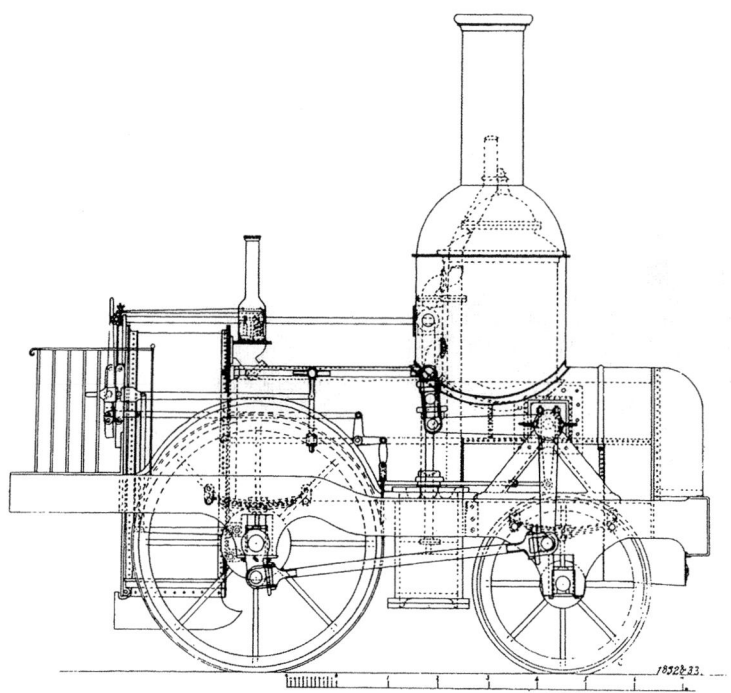

Roberts' first locomotive, the *Experiment*, was delivered to the Liverpool & Manchester Railway in May 1833. It was not a success for it consumed 40 per cent more coke than the other engines, the variable expansion valves did not work properly and the vertical cylinders made the engine ride roughly

1832. The *Experiment* was a 2-2-0 tender engine, driving wheels of 5 ft diameter, front wheels 3 ft 2 ins, and total weight 8¾ tons. To avoid the weakness of a crank axle, the cylinders were placed vertically either side of the engine between the wheels, driving through a bell crank which gave a very rough ride. The boiler design was unconventional as well, with the chimney placed over the dome in the middle. The *Experiment* had a high fuel consumption and was quickly sold to the Grand Junction Railway in 1836. While the next three locomotives for the Dublin and Kingstown Railway in Ireland had conventional boilers, Roberts placed the cylinders vertically over the front carrying wheels, driving through bell cranks and even shorter connecting rods. These alterations probably made these locomotives even more unstable because their front ends were reputed to oscillate up and down violently at any speed, which caused the rails to break.

Sharp, Roberts's standard railway locomotives

Roberts was quick to learn from his mistakes, or perhaps it was the arrival in the drawing office of a young German, Charles Beyer, who improved the design of the next order for Sharp, Roberts's locomotives. The ten 2-2-2 tender engines ordered by the Grand Junction Railway in November 1835 were prototypes for the standard Sharp, Roberts's locomotives with this wheelbase. They had cylinders 12½ ins in diameter by 18 ins stroke, driving wheels 5 ft in diameter, carrying wheels 3 ft 6 ins in diameter, weight 12 tons 5 cwt. While the diameter of the driving wheels was increased to 5 ft 6 ins for later orders, the other dimensions did not alter much. Roberts had designed a standard type of locomotive which could be built to standard templates. Between 1833 and when Roberts left the firm in 1843, 79 orders were received for 246 locomotives.

In 1837, 11 orders were received for about 50 locomotives and once again it must have become clear that the Globe Works did not have the capacity to match demand. T. Sharp's iron business was located alongside the Rochdale Canal by Oxford Street and this site was chosen to be developed into the Atlas Works specifically for the construction of railway locomotives. Roberts would have supervised most of the design. A range of new multi-storey workshops and a foundry were constructed in 1839 at a cost of £30,000. On Roberts's death in 1864, it was said that the Atlas Works were still ranked 'among the largest and most productive engineering factories in the kingdom'.[11]

Roberts applied the same techniques to the construction of his locomotives as he had done to textile machines. He had templates made so that batches of the same parts would all be alike. He developed special machine tools for particular tasks, such as a slotting drill for cutting the oil grooves in bearing brasses. He had rotary cutters

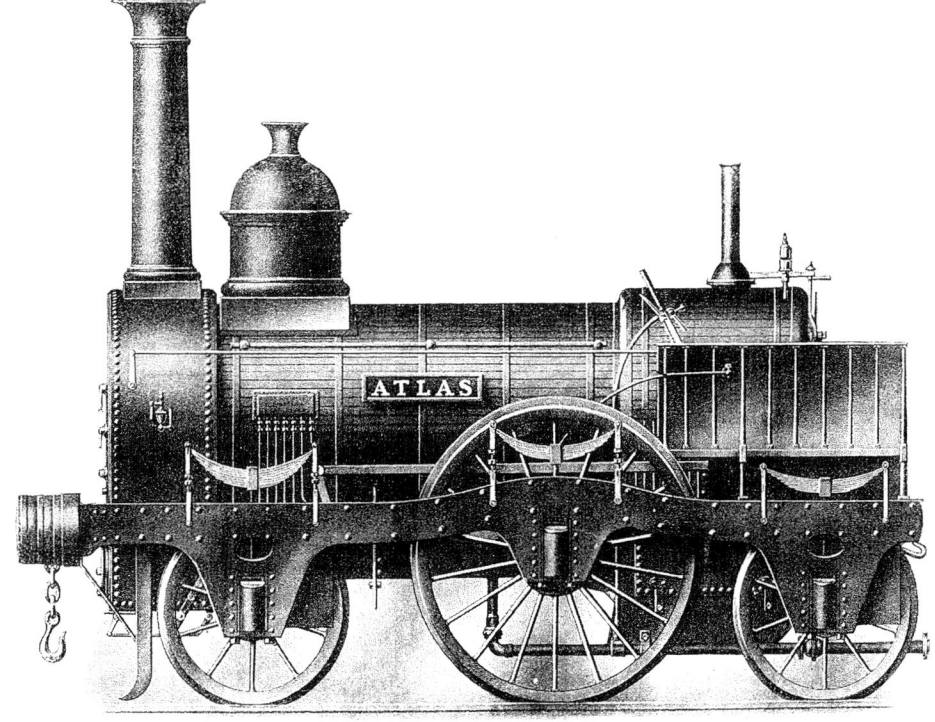

A typical Sharp Roberts 2–2–2 tender locomotive built about 1838

for finishing the heads of bolts or carving out crankshafts. He must have had a production line of tools for making bolts which could turn out 1000 a day. He improved punching and shearing machines so that they were 'considerably simplified, and rendered capable of doing 50 per cent more work and of better quality'.[12] Roberts used punching machines to cut out the curving main frames for his locomotives as well as punch out rivet holes on boiler plates. He invented the method of shaping the curved plates for boiler barrels by passing them through a set of three rollers in which the height of the centre one could be adjusted to give the correct curvature. The result was that '[t]he engines built by Mr. Roberts were excellent in design, compact in form, capable of doing hard work, and doing it well'.[13] This was the most successful era of Roberts's career.

Personal life

In the meantime, Roberts must have married for a second time around 1832 and had three children, John born in 1833; Eliza Mary born in 1835 and a second son William born in 1843. In 1838 he was elected a member of the Institution of Civil Engineers, an honour of which he was proud. In the same year, he was elected Councillor for the Oxford ward of Manchester, which he continued to represent until 1843. In March 1839, he was involved with founding the Royal Victoria Gallery for the Encouragement and Illustration of Practical Science, which was intended to arouse the interest particularly of the young through demonstrations and exhibits. William Sturgeon was appointed as director but the venture failed in 1842.

In 1835, Roberts attended the meeting of the British Association for the Advancement of Science in Dublin. He went to many others later, including that in Manchester in 1842. He took the opportunity to display many of his smaller inventions at these meetings. Around 1836 or 1838, he invented what he called a Centrifugal Railway, where a truck filled with water looped the loop without spilling any. Larger versions capable of carrying a person were built in the Manchester Hall of Science and some pleasure gardens. Roberts had branched out into the manufacture of turret clock mechanisms by 1838 and his first was placed on the Atlas Works where it became well known to Mancunians. Sturgeon may have stimulated Roberts's interest in electro-magnetism so that he constructed several large electro-magnets, one of which would sustain a load of 1,400 lbs.

No reasons have been found for the break-up of the firm of Sharp, Roberts in 1843. Roberts retained the Globe Works which he carried on alone for 18 months but became ill. He then formed a partnership, with Benjamin Fothergill superintending the works and

Robert Graham Dobinson taking charge of the accounts. Fothergill retired through ill health in 1849 or 1850. But the concern was not profitable and was closed down at the end of 1851. The machinery and contents were auctioned in February 1852 and the business finally wound up in June 1852. Roberts became a consulting engineer, retaining an office at the Globe Works until probably some time in 1857 when he moved to 30 Brown Street where he remained until he left for London in the summer of 1860. Neither Fothergill nor Dobinson could provide the financial backing for Roberts's new inventions as Hill and Thomas Sharp had. The failure may also have been due to Roberts's lack of business acumen since the range of his subsequent patents suggest that he spread his net too wide.

Later inventions

It has been suggested that Roberts wished to concentrate on his own inventions. There are 25 patents in his name. Between 1843 and 1851 he took out seven and from then until 1860 there were 11. Some cover a multitude of claims for novel inventions, among them all sorts of improvements to textile machines, covering cotton gins, opening machines, combing machines, the mule, looms for weaving velvet, finishing machines and a machine for engraving printing plates or rollers. There were improvements to clocks and watches, some of which were displayed at the Crystal Palace Exhibition in 1851 where he was awarded Prize Medals for his Alpha turret clock mechanism and a tool for drilling watch plates accurately. Water turbines, meters for liquids, metal storage casks and many more ideas featured in his patents. We know that prototypes of many were built because they were listed in the 1852 sale. In addition, Roberts said that he had over 100 inventions which he never patented. Some were his improvements to an American cigar-rolling machine.

Roberts won a prize medal for his 'Alpha' clock mechanism at the Crystal Palace Exhibition in 1851

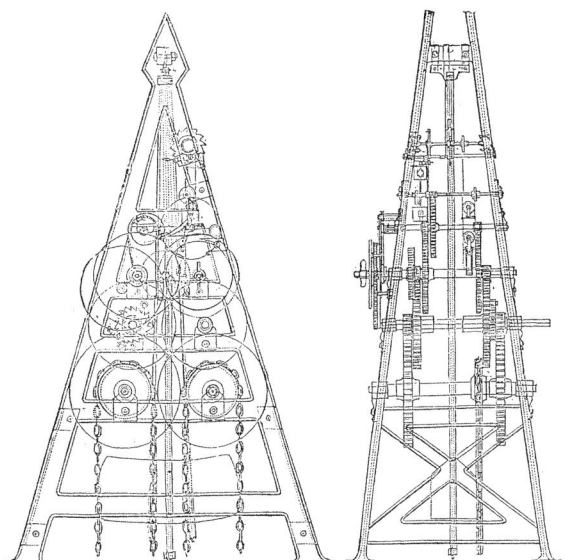

It is doubtful whether any of these later patents were profitable. In the 1820s, Roberts was pioneering production engineering alone. After the 1830s, he faced competition from people like Joseph Whitworth, James Nasmyth and a host of others.

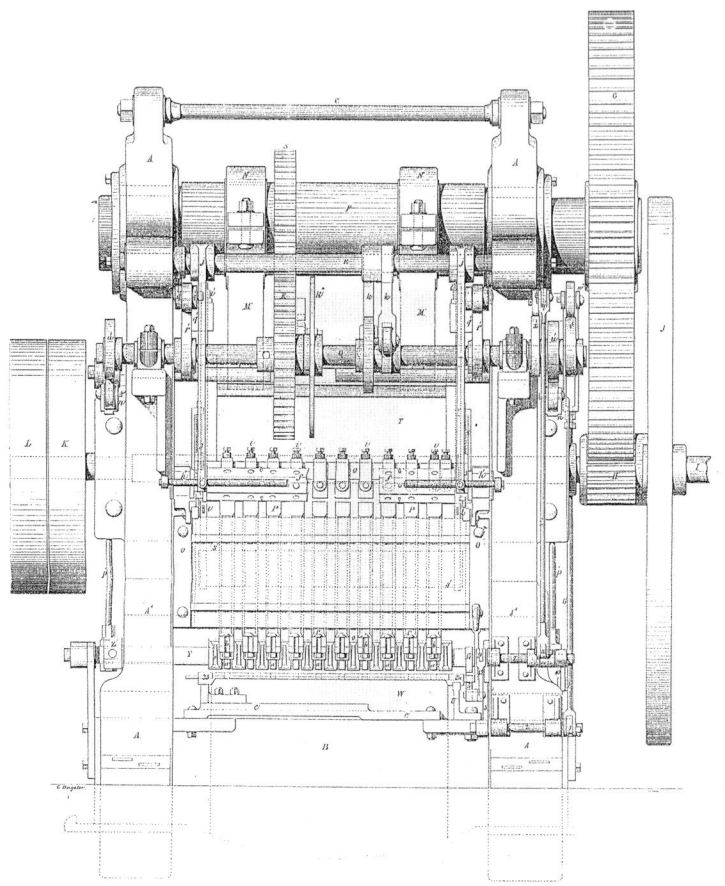

Rear view of the Jacquard punching machine. There are ten punches near the bottom, with the cylinder (shown broken in the middle) for the Jacquard cards near the centre

But Roberts did make one famous invention, the Jacquard plate-punching machine. The story, as told by Smiles, is that the contractor building the Conway tubular railway bridge was afraid that the first tube would not be completed on time because the men were 'going slow' and he wanted a quicker way of punching the rivet holes in the iron plates. Roberts was approached and he evolved his Jacquard machine while quietly sipping his evening tea. Construction of the first tube did not start until the summer of 1847 but the date of the application for the patent is before that, on 5 March.[14] It was a very complex machine in which any of a row of ten punches could be selected by a Jacquard apparatus as the plate moved through, one space at a time. One mechanic, with three labourers to lift the plates on and off and a boy to oil the punches, could punch a 12 foot-long plate with 144 holes in four minutes more accurately and quicker than 12 men on earlier machines. While Roberts's machine saved the day on the Conway bridge and his machine was used on other similar bridges, only two were ever built. Roberts may have designed it for punching holes in boiler plates, but it was unsuitable for this because it could only take flat plates and the holes would have been distorted through subsequent bending.

The structure of the Conway bridge gave Roberts the inspiration that iron ships, then in their infancy, should be built on similar tubular and cellular principles for strength. He took out a patent in 1852[15] with 33 new claims shown on five sheets of drawings. He had designed a passenger liner with accommodation for 500 passengers. Had it been built, it would have been the largest ship afloat, eclipsing Brunel's *Great Britain*. Practically every aspect of ship design was covered, from the structure, ventilation, improved boilers, coal bunkers loaded by trucks on railways, expansive steam engines for greater economy, twin screws for manoeuvrability, even better life-boats and naval armaments. Some of these inventions were well ahead of their time.

Most nautical people ridiculed his ideas, which he tried to promote at meetings of the British Association, the Institution of Civil Engineers and even with the French Emperor. His ship became an obsession on which he spent a large fortune but found no backers. It was the cause of his leaving Manchester in 1860 so that he could defend his ideas since Fairbairn had proposed similar ones in a lecture in March 1860 and was claiming the credit. In London, Roberts was looked after by his daughter and set up an office as a consulting engineer at 10 Adelphi Street, just round the corner from the Institution of Naval Architects, of which he was an associate. Although at least one ship, the *Flora*, was fitted with twin screws, Roberts made little progress with his other ideas.

In 1861 it was discovered that he was living in straitened circumstances and an appeal was launched to raise funds to support him. A civil list pension was turned down by the Prime Minister, Palmerston, on the grounds that the amount available was very small and that Roberts's case did not fall within the rules for making a grant. Palmerston also pointed out that the manufacturers who had profited from Roberts's inventions could well give a better annuity than the civil list. But before help could be given, Roberts, inventing to the end, died in his daughter's arms on the morning of 11 March 1864 partly as a result of a fall down some steep stairs. The obituary in *The Engineer* stated,

> One of the most gifted inventors and skilful mechanical engineers, who within the last half century, have added so much not only to the wealth of England but of the whole civilized world, has just passed away, at the ripe age of seventy-five. It is not too much to say that the inventions of Richard Roberts rank in value with those of Arkwright, while for exquisite ingenuity, his self-acting spinning

mule – still worked in many cases in almost the form in which it was introduced to quell the Staleybridge riots – surpasses any description of mechanism known in modern manufactures.[16]

Roberts's final poverty may have been due to a lack of business acumen. The partnership with Thomas Sharp was not replaced with a Matthew Boulton or an Hon. C. S. Rolls to provide the necessary capital or supervision. But even if Roberts died in reduced circumstances, Manchester certainly benefited from his passion for invention, in particular for machine tools, which enabled the city to retain pre-eminence in manufacturing industry for so much of the nineteenth century.

Notes

1. For a fuller biography, see R. L. Hills, *The life and inventions of Richard Roberts, 1789–1864* (Ashbourne, 2002).
2. W. Pole (ed.), *The life of Sir William Fairbairn* (London, 1887; reprinted Newton Abbot, 1970), p. 112.
3. Birmingham Central Library, Matthew Boulton Papers 464, T. J. Wilkinson to M. R. Boulton, 25 Feb. 1823.
4. Patent 4,726, 14 Nov. 1822.
5. Patent 5,138, 25 Mar. 1825.
6. Patent 5,949, 1 July 1830.
7. R. Marsden, *Cotton spinning* (London, 4th. ed., 1891), p. 228.
8. W. Johnson, *The imperial cyclopaedia of machinery* (Glasgow, n.d.), p. 57.
9. *The Engineer*, 17 (1864), p. 175.
10. Patent 6,528, 13 Apr. 1832.
11. *Minutes of the Proceedings of the Institution of Civil Engineers*, 24 (1864–5), p. 539.
12. *Ibid.*, 17 (1857–8), p. 194.
13. *The Engineer*, 17 (1864), p. 184.
14. Patent 11,607, 5 Mar. 1847.
15. Patent 14130, 22 May 1852.
16. *The Engineer*, 17 (18 Mar. 1864), p. 175.

Cosmos, climate and culture: Manchester meteorology made universal

Graeme Gooday[1]

This article draws a historical connection between Victorian meteorological investigations in north-west England and the wider operations of the universe. It shows how, in the second half of the nineteenth century, a number of Mancunians helped to forge a radically new form of 'cosmical meteorology'[2] on just that basis. The foremost among these was Joseph Baxendell (1815–1887), astronomer to the Manchester Corporation, 1857–1871, and to the Southport Corporation thereafter. Two other major characters were both physicists at Owens College, Manchester: Balfour Stewart (1828–1887) and Arthur Schuster (1851–1934). All three were linked through their involvement with the Manchester Literary and Philosophical Society (MLPS), at which each presented important research on the relationship between meteorology, terrestrial magnetism and sunspots, much of it subsequently published in the MLPS *Proceedings* or *Memoirs*.[3] I show how Baxendell and Stewart promoted cosmical meteorology through this important regional periodical, as well as other local and national media from the 1860s until their coincident deaths in 1887. I conclude with a brief discussion of how Schuster sustained this much-criticized project in a less ambitious form at what later became the University of Manchester.[4]

Joseph Baxendell, corporate astronomy and weather troubles

From the point of view of history of science and culture in the Manchester region, Joseph Baxendell is an important and interesting character. Although his adult career was located in the north west his meteorological and astronomical work was of both national and international significance. Born on 19 April 1815 at Bank Top, Manchester, Baxendell attended Thomas Whalley's school at Cheetham Hill but was largely self-educated thereafter in mathematical and observational sciences. Both biographical and institutional forces drew him to correlated researches in meteorology and astronomy. From the ages of 14 to 20 he served as an ordinary sailor on the *Mary Scott* in trading trips to South America, gaining the mariner's intimate appreciation of weather patterns. It was apparently some 'casual remarks from his officers on matters of navigation' that inspired him to study the natural world, and thus as a teenager to develop the skills in observation for which he became renowned.[5]

Baxendell returned to Manchester in 1835, first assisting his father as a land steward before developing an independent business as an estate agent. He habitually dedicated his spare hours to astronomy, especially assisting his friend Robert Worthington (who had been accidentally blinded in one eye) in operating a five-inch refracting telescope at the latter's private observatory at Crumpsall Hall, near Manchester. By identifying and studying as many as 18 hitherto unknown variable stars in ensuing decades, Baxendell eventually became one of Britain's pre-eminent systematic observers in the field.[6] After a decade of nightly telescopic work the retiring autodidact finally took his work into a public forum by submitting his observations on ë Tauri[7] to the Royal Astronomical Society (RAS). Nearly another decade passed before Baxendell published again, four of his papers on variable stars appearing in the *Monthly Notices* of the RAS between 1855 and 1857; on the strength of these Baxendell was elected a Fellow of the RAS in 1858.[8] In that year he was also elected a Fellow of the MLPS, having recently presented further astronomical findings on variable stars in the series of MLPS *Proceedings* launched in 1860.[9] With his new credentials as a nationally and locally acknowledged observer of the heavens, he was appointed Astronomer to the Corporation of Manchester in 1859. As a post carrying an honorarium of a mere £40 a year – considerably less than an ordinary labourer's annual earnings – its holder was presumptively a gentleman specialist of independent means.

To understand why this post existed at all, though, we need to consider the local and national context of Manchester as a city increasingly connected to the rest of Britain via a burgeoning rail network. In 1847 the City of Manchester Corporation took upon itself a duty to provide civic notification of London Greenwich time, thereby displacing Manchester 'solar' time as the reference for local business and railway transactions. The traditional profusion of different local times across the British Isles generated not only dangers for the minute scheduling of railway signalling, but also a confusion over timetables that frequently led passengers to miss trains. Five years after the Great Western Railway introduced standard London time for all its services in 1840, the Liverpool and Manchester Railway Company petitioned Parliament for a single uniform time to be imposed on all businesses. This was not legally implemented until 1880 and indeed it was initially strictly voluntary action by the railway companies that brought London time to Britain in general and Manchester in particular. The

North Western Railway introduced Greenwich time to its stations at Liverpool and Manchester in January 1846, and two years later this had been adopted by most railway companies in the north-west region.[10]

Thus it was in 1847 that the City Council of Manchester resolved to expend £100 to purchase a clock – along with certain unspecified meteorological instruments – and 'place them in an accessible position in the Town Hall … for the purpose of informing the public at all times what is Greenwich time'; all other clocks under the Corporation's control were set to and regulated by this clock. Initially this clock was placed under the care of Manchester's best known clockmaker, Peter Clare (1781–1851), a mathematician, anti-slavery campaigner and close associate of Joseph Dalton. Mancunians evidently persisted in using solar reckoning, however: when 'Quaker Clare' was offered declarations of solar time he would hastily reply: 'Dost not thee know that the sun has been notoriously wrong a long while, and that nobody minds him now? I'll tell thee what the time is by my clocks.' After Clare's death, the Council decided in 1852 to opt instead for astronomical time-keeping methods and spent £120 on a transit telescope, sidereal clock and chronometer for the purpose.

The first Astronomer to the Corporation was the Baptist minister Reverend Henry Halford Jones, who had long contributed astronomical information to Manchester's almanacs and occasional astronomical papers to the MLPS and RAS (of which he was member and Fellow respectively). Installing the instruments in an observatory next to his house, Jones kept Manchester's clocks regulated with unprecedented precision by using a pendulum device that needed little compensation for weather conditions; as a tacit perk of the position, he was also able to use the transit telescope to undertake his own private astronomical research.[11] After Jones's death in late 1858 the post of Astronomer to the Corporation naturally fell to another Manchester FRAS, Baxendell inheriting Jones's astronomical duties and instruments in February 1859.

From several other points of view, the year 1859 was a turning point in Baxendell's career, turning it from private observer to civic sage. It was then that he became MLPS secretary, serving in that role until 1885, and serving as editor of its *Memoirs* till his death. Most importantly for our purposes, it was the year in which Baxendell helped to found a specialist 'Manchester Literary and Philosophical Society – Physical and Mathematical Section' within the MLPS. During the winter months of the years that followed, it was to this small erudite body – numbering half a dozen – that Baxendell presented his astronomical and meteorological observations, notably those that Balfour Stewart judged to be of crucial importance in linking the periodic variations of sunspots and terrestrial weather. While Baxendell's attention to such cyclic phenomena was undoubtedly primed by decades of studying periodic behaviour in remote stars, what was it that prompted him to reconnect meteorology and astronomy some 24 years after he had left the marine service?

Two major events in the autumn of 1859 inspired British attempts to develop a 'science' of meteorology, long treated as a natural history of collecting climactic facts and figures. A huge solar flare on 1–2 September 1859 brought, as we shall later see, magnetic storms that had extraordinary meteorological and electrical consequences around the world. More catastrophically, however, on 25–26 October that year a powerful tempest brought chaos to much of Britain, causing widespread destruction of railways and buildings. Most specifically it produced one of the most savage storms ever to strike the Irish Sea, wrecking the steamer *Royal Charter* on the Isle of Anglesey, just off the north Welsh coast. The ensuing loss of 459 lives and a huge cargo of gold *en route* from Melbourne to Liverpool shocked even the most hardened Victorian stoic. It prompted Rear-Admiral Robert FitzRoy to launch plans he had long nurtured for a system of what he was the first to call weather 'forecasts'. Two years later Fitzroy's cautionary storm warnings were not only sent to the Admiralty, ports and insurance companies, but published in national newspapers too.

Balfour Stewart, Professor of natural Philosophy at Owens College, Manchester, 1870–87

Yet the storm-warning services did not bring Fitzroy great credit among his employers at the Board of Trade's meteorological board. They eventually instructed him to cease his controversial forecasting as it was beyond the bounds of his duties as they conceived them. Already suffering from difficult personal circumstances, Fitzroy committed suicide in April 1865 rather than witness the end of his life-saving work.[12] As an erstwhile sailor, Baxendell responded in sympathy by publishing a private pamphlet (drawn from a paper he had read to the MLPS) vigorously protesting against cancellation of the storm warnings.[13] When the Royal Society took over the supervision of the meteorological department six years later it refused to deal with forecasts, and indeed Baxendell's hostility to its policies probably explained his comparatively late election as a Fellow of the Royal Society in 1884.[14]

During the 1860s, and until his departure to Southport in 1871, Baxendell's commitment to utilitarian causes was manifest in many areas beyond that of storm warnings. A major concern was to determine the influence of meteorological conditions on public health and communicate the anticipated consequences to civic authorities. For example, his warning of an impending dry summer in 1868 assisted the Manchester Corporation Water Works to regulate water supply so as to pre-empt the worst consequences of drought.[15] Moreover, Baxendell took his astronomical duties as being more than just the regulation of Mancunian clocks to Greenwich Time; this was in any case a sinecure once railway expansion had brought an effectively instantaneous telegraph link between London and the north west. His long term aim was to enhance weather forecasting by using his astronomical skills to identify cyclical correlations between solar phenomena and terrestrial meteorology.

Baxendell and the development of astro-meteorology

During the 1860s, Baxendell's presentations to the MLPS ranged very broadly in their cosmic endeavours. He sought to identify correlations between temperature and pressure in the seasonal variations of Europe and Asia, later in the decade attempting to find links to the moon's orbit and to variations in the earth's magnetic field. For example, in a paper communicated to the Physical and Mathematical Section on 5 March 1863 he noted from measurement data at Greenwich in the period 1848–60 that the pattern of variation in solar radiation could be correlated to records of terrestrial temperature if the pattern of solar spot frequency was factored in to explain the discrepancies.[16] Lured by publications in the RAS *Monthly Notices* to study the periodical variations of sunspots (as a possible correlate of stellar variability),[17] his plan was to make a more systematic study of whether variation in the number and size of the sun's (cooler) surface spots could explain the variation in terrestrial temperatures at Greenwich and elsewhere. This project generated considerable interest in the winter of 1867–8, and later, especially when published by the MPLS in 1871.

On 10 October 1867 Baxendell read the first part of a paper 'On solar radiation' to the MLPS Section.[18] He revealed the enormous difficulties of attempting to draw together data on solar activity gathered by different observers using black-bulb thermometers that he needed to fulfil his long-term ambitions for weather forecasting. Of the 'perplexingly anomalous and unsatisfactory' results he had obtained from Greenwich he related that:

> Although observations of solar radiation have now been regularly made for several years at various public observatories, and by many amateur meteorologists, I am not aware that any useful or important result has yet been deduced from them. It seems to be generally supposed that the disturbing influences which affect the indications of the black-bulb thermometer are so uncertain and irregular in their action as to render it almost hopeless to expect that any new and valuable result can be obtained from them. On comparing sets of observations made by different observers, the most startling, and discouraging discrepancies are often found to exist, for which, in the absence of any information as to the exact circumstances under which observations were made, it is impossible to account satisfactorily.[19]

Fortunately Baxendell had recently been given volumes of observations of solar radiation and temperature at Oxford for 1858–64 by the University's Radcliffe Observer, Revd Robert Main FRS. Although he had not completed his analysis of the correlation for this data set, he felt that his results were 'sufficiently curious and remarkable' that he should draw them immediately to the attention of fellow researchers, albeit with a view to devising and adopting 'more reliable and systematic methods'. Showing the temperature and solar radiation figures in graphical form next to observations of sunspots for the same period revealed two curves of 'strikingly' similar form. For Baxendell this was 'conclusive' evidence that a connection existed between the two classes of the phenomena.[20]

Not all in Baxendell's expert audience in the MLPS Physical and Mathematical Section were able completely to confirm his results. One such was the section's secretary, the cotton-spinner George Vernon, like Baxendell an FRAS but also a Fellow of the Meteorological Society. On 26 November he reported that his own 11-year pattern of solar radiation observations made at Old Trafford produced a distribution that did not 'quite agree' with Baxendell's. Vernon did at least confirm the general correlation adding, in bucolic mood, that the greatest relative effect of solar radiation was in the spring in accordance with the 'very rapid growth of vegetation' that Manchester saw during that season.[21]

On New Year's Eve 1867 the Revd Thomas Mackereth FRAS, FMS, a schoolmaster in Salford and resident of Eccles, offered two papers to the Physical and Mathematical Section, explicitly inspired by Baxendell's work and indeed assisted by him. The first of these investigated the reliability of methods employed by the Oxford observers in his presentation of 'A comparison of solar radiation on the grass and at six feet from the ground'. As the Corporation of the Borough of Salford had recently installed a meteorological station in front of its Town Hall he used the roof apex of its shade-stand six feet above ground as an ideal location for his solar radiation thermometer; for a comparative study he fixed a similar thermometer in a comparable position at Eccles. At first he was most struck by the discrepancy in results between the elevated thermometers and those on nearby grass, but achieved much better correlations using a thermometer placed *in vacuo* on the grass. Mackereth thus concurred with Baxendell that 'some definite principle' was required in the placing of solar thermometers to attain any rigorous results.[22]

Mackereth's second paper presented solar radiation observations that Baxendell had encouraged him to make at Eccles, especially given that Mackereth's thermometers had long been set up much like those in Oxford. Although presenting his results in the same format as Vernon, Mackereth found his results for Eccles matched Baxendell's from Oxford rather more closely than Vernon's. Given his close association with Baxendell, it is perhaps not surprising that Revd Mackereth anticipated the match would have been 'more striking' still had he extended his observations over a longer period.[23]

A month later, on 28 January 1868, Baxendell read the second part of his paper 'On solar radiation' in which he more thoroughly presented his case for a connection between solar radiation and terrestrial temperature, making various seasonal corrections for the transmission of sunlight through water vapour in the atmosphere. He was now also able to assimilate Mackereth's results for Eccles, extending as they so usefully did two years beyond the Oxford series. Baxendell inferred from the results:

> It appears, therefore, that the calorific intensity of the sun's rays continued to diminish for two years after the termination of the Oxford series; and as the observations of Schwabe, Wolf, Balfour Stewart and others have shown that the frequency of solar spots also diminished during these two years, the probability that a close connexion exists between the two phenomena is considerably increased by the results of Mr Mackereth's short but valuable series of observations.

Once again a key feature of Baxendell's argument was to show closely congruent curves between the two sets of data, thus using the weather at Eccles and Oxford to generalize at a cosmic level about the processes linking solar spots to terrestrial effects of solar radiation.[24]

Published by the MPLS in 1871, this paper made a great impact on the newly arrived Professor of Natural Philosophy at Owens College,

Manchester, Balfour Stewart – and not just for reasons of collegial mutual citation. As Stewart said in his obituary of Baxendell for *Nature* in October 1887, notwithstanding the errors of detail:

> Baxendell's contributions to meteorology are very important, and in one branch of this science, he may claim to be the pioneer. In 1871, from an analysis of eleven [*sic*] years' observations of the Radcliffe Observatory, Oxford, he came to the conclusion that the forces which produce the movements of the atmosphere are more energetic in years of maximum than in years of minimum sunspot activity. This conclusion has since been confirmed by other observers.[25]

Stewart had to admit, of course, that Baxendell's work was highly contentious since not all observers had been able to confirm his conclusions. Indeed, hinting at the carping of Richard Proctor, Stewart reported that:

> We have heard it objected that Baxendell generalized from a comparatively small number of observations, but in a question like this such a procedure is essential to the pioneer. His task is to deduce with a mixture of boldness and prudence something of human interest out of the mass of observations already accumulated, and thus to stimulate meteorologists not only to go on with their labour, but to cover more ground in the future than they have covered in the past. Baxendell's procedure in this respect has been abundantly justified by the fact that many other men of science are now following in his footsteps.

As we shall see, Stewart himself was one of those who followed in Baxendell's footsteps, having already become a major scholar on the theory and observation of sunspots before he arrived at Manchester in 1870.

Balfour Stewart: taking heavy weather from Kew Observatory to Owens College

Although more of a high profile public figure than Baxendell, the Edinburgh-born and educated Balfour Stewart has won relatively little attention from historians for his meteorological work or role in Manchester science.[26] Yet as Superintendent of Kew Observatory in London, 1859–70, and as Professor of Natural Philosophy at Owens College, Manchester, 1870–87, he was widely held as an authority on both astronomy and meteorology. Stewart enjoyed particularly close associations with Norman Lockyer (founding editor of *Nature*), Peter Guthrie Tait (Professor of Natural Philosophy at Edinburgh University)

Arthur Schuster, Langworthy Professor of physics at Owens College, 1881–1907

and Arthur Schuster who was student (1871–3), assistant (1875) then colleague at Owens College, taking up the Langworthy Professorship of Physics from 1881. Like Baxendell, however, he often took on the scientific establishment in principled conflicts that had negative consequences for his career; indeed Stewart generally fared ill in his conflicts with patriarchal Sir Edward Sabine (President of the Royal Society), and with uncooperative field-station observers around Britain. Most troublesome perhaps were Richard Proctor and Robert H. Scott of the Meteorological Office, who both repeatedly challenged claims by Stewart and his allies that they had identified close connections between terrestrial weather and solar turbulence.

Two months before the wrecking of the *Royal Charter*, and rather less lethal, was the globally visible spectacle of a huge solar flare on 1–2 September 1859. This generated a magnetic storm so intense that aurorae normally seen only at the poles were unusually visible near Rome and Hawaii. Telegraph lines around the world went haywire, garbling messages, giving electric shocks to signal-men across the telegraph network and setting fire to telegraphic apparatus in Norway.[27] Occurring just after Stewart arrived at Kew Observatory in southwest London, he was well equipped to observe and record precisely how major changes on the sun's surface could have a sudden and substantial effect on terrestrial life.

Kew Observatory was well placed to monitor this relationship since Stewart's boss there, now Major Edward Sabine, had for some years been convinced that sunspots were linked to changes in the earth's magnetism. The major difficulty in exploring this clue to solar influence on terrestrial conditions was that astronomers could not agree whether it took ten years or 11.11 years for sunspots to circulate around the sun's surface.[28] The period of the sunspot cycle long remained a point of contention, especially in later speculations that Stewart sought to make about links between sunspots and the weather. In addition to calibrating thousands of thermometers and barometers sent to him from field-stations around the country, from 1862 Stewart pursued Sabine's programme of solar meteorology, monitoring the motion and size of sunspots with the photoheliograph installed at Kew for the purpose of mapping the weather on the sun's surface. Two years later Stewart identified a possible explanation for sunspot cycles in planetary influence: sunspots were generated at points of the sun in opposition to nearby planets Venus and Mercury, and grew in size as the planet orbited away from them.[29] Such speculative claims were queried sharply by Richard Proctor, however.[30]

The increasingly ambitious Stewart next turned Kew into a laboratory for studying how the planets might interact with the sun. From 1865 he collaborated with Peter Guthrie Tait, a fellow Edinburgh graduate, in studying how discs spin to a halt in a vacuum. They claimed to have found friction in the cosmic ether – key to understanding how energy could move across empty space between planets and the sun. Others such as George Gabriel Stokes, Secretary of the Royal Society, saw rather more mundane sources of friction at work. Trials were continued in the 1870s, probably with Stewart's students at Manchester, involving the youthful J. J. Thomson, but never achieved conclusive results.[31]

Despite the controversies, Stewart began to collaborate in 1868 with scientific journalist and amateur astronomer Norman Lockyer to popularize the connection that he claimed must exist between sunspots and earthly weather. In the article they co-authored for *Macmillans Magazine*[32] they tellingly did not use Stewart's data from Kew Observatory. Rather they relied on the 'unconfirmed' testimony of an observer in Manchester – Joseph Baxendell – that terrestrial temperatures were affected by sunspots.[33] Combining this with evidence of concurrent planetary influences, they moved quickly to conclude that:

> the different members of our [solar] system are more closely bound together than has been hitherto supposed. Mutual relations of a mathematical nature we were aware of before, but the connexion seems to be much more intimate than this – they feel, they throb together ... [and] something of this kind might be expected if we suppose that a Supreme Intelligence ... pervades the universe, exercising a directive energy capable of comparison with that which is exercised by a living being.[34]

Stewart seems to have made this grandiose claim for a divinely unified meteorology of the cosmos to win support for Kew Observatory to become the programme's central body. Since the previous year his prospects of achieving this aim had been enhanced by the government's restoration of its Meteorological Committee to quell the controversy that followed its abandonment of 'storm warnings'.

Stewart was appointed Secretary to this Committee, and secured for Kew the status of 'Central Meteorological Observing Station'. From this vantage point he could co-ordinate the results of the weather stations around the British Isles with the results of solar observations. Yet Stewart's heavy-handed attempts to refashion the Meteorological Committee's system of weather surveillance to meet his 'cosmical' agenda soon resulted in disastrous conflict with Sabine.

As we know from the later testimony of Stewart's junior colleague at Manchester, Arthur Schuster, in his *Biographical fragments*, Sabine saw Kew's facilities as serving other purposes.[35] By 1869 Stewart readily had recourse to the columns of Lockyer's new journal *Nature* to complain about his treatment by Sabine. There he bewailed how the scientific worker had to 'work with the one hand and fight with the other', and railed against the 'deplorable' lack of co-operation and systematic practices among dispersed observers.[36] By 1870 his bid to marshal Kew's resources into a scheme of cosmical meteorology was untenable and he (effectively) resigned his Superintendency by taking up the Chair of Natural Philosophy at Owens College, Manchester.[37]

Owens College and *The Unseen Universe*

Although Manchester had almost none of Kew's technological advantages, it at least offered Stewart direct communication with Baxendell through meetings of the MLPS Physical and Mathematical Section. It also offered Stewart a forum for research free from Sabine's constraints and a captive student audience that could serve as a workforce for him. In his inaugural address at Owens College, Stewart declared that:

> [I]t is of great importance to know whether the earth's climate and atmosphere are influenced in any way by the changes taking place in the atmosphere of the sun. Such a connection has not yet been traced, but it has hardly been sought for in a proper manner ... I feel convinced that meteorology should be pursued in connection with terrestrial magnetism and solar observations; and were a well considered scheme for solving this great problem fairly introduced, I am sure that scientific institutions and individuals throughout the country would do all that they possibly could do to promote this most important branch of physical research.[38]

Stewart's researches and speculations were prominent, especially his campaign for a new publicly-funded observatory – apparently to be run by himself – conducted both in interviews with the Devonshire Commission[39] (Secretary, Norman Lockyer) and in polemical articles for *Nature* (editor, Norman Lockyer). Yet despite the popular interest during the 1870s in both sunspots[40] and weather forecasting,[41] these stage-managed demands came to nothing. A similar fate befell his pleas – akin to Baxendell's – that all weather data be published unprocessed so that his universalizing techniques for reducing observations could transcend the local idiosyncrasies of other observers' programmes.[42] So instead of a single 'well considered scheme', co-ordinated from his Owens College laboratory observatory, there emerged a multiplicity of localized efforts around the world, each with their own divergent agendas, resources and techniques.

Cultural anxieties about and responses to weather have long been appropriated by the ambitious;[43] indeed sunspot-weather linkages could be 'appropriated' for a number of rather different concerns. British colonial administrators, for example, had long been preoccupied with famine and flood relief;[44] so after identifying a similar 11-year cycle in the monsoon rains of Ceylon (Sri Lanka), Lockyer tried to persuade the Indian Meteorological Department (opened in 1875) that India's regular climatic variations could easily be correlated with sunspot cycles. This would provide him with financial support for his own laboratory and a chance to prove the utility of his solar work.[45] For Stewart and Tait, the linkages could even be harnessed to 'prove' the compatibility of science and religion, as was claimed in a treatise they published in 1875 (initially) anonymously entitled *The unseen universe*. Here is how a reviewer in *Nature* summarized one of its arguments:

> a whole series of tremendous meteorological phenomena, such as hurricanes in the Indian Ocean, happen because certain positions of Mercury and Venus affect the sun's atmosphere, causing spots ... and th[is] condition of the sun affects the earth. Like the complicated series of effects which follow the pulling of the trigger of a gun, the effects are utterly disproportionate to their causes. Man is a machine of this unstable kind ... May not other beings [thus] be capable of touching what we may call the hair-triggers of the universe? Whatever these agencies are, angels or ministering spirits, they certainly do not belong to the present visible universe.[46]

Significantly, the work of Scottish natural philosophers was cited as evidence on the management of weather by cosmic spirits in the ether, especially Stewart and Tait on discs rotating *in vacua*![47] In fact their co-authorship was obvious prior to confirmation in 1876 in the fourth of the 14 editions that appeared over 13 years.[48] The critical debate around this much-read book was intense;[49] yet while reviewers were generally sceptical of its arguments – the *Nature* reviewer doubting that the 'invisible universe' could be supported eternally by energy *dissipated* in the 'visible' – none challenged evidence of a link between sunspots and weather.[50] Other than to confirm Stewart and Tait's indefatigable piety, the only unequivocal effect of *The unseen universe* was thus perhaps to give publicity to cosmic meteorology – but not enough, as it turns out, to secure its universal credibility.

The disintegration of universal meteorology from the late 1870s

Within a few years, the recurrent proliferation of independent investigations had produced troublesome disagreements. While in some locations a positive correlation was found between cycles of sunspot intensity and key meteorological parameters, for others precisely the *opposite* correlation was claimed: Balfour Stewart admitted this of Meldrum's work on rainfall at a meeting of the Manchester Literary and Philosophical Society in 1880 (see below). Worse than this, at about the same time, the Indian Government's chief meteorologist announced he could not verify Lockyer's claim that droughts regularly followed sunspot minima.[51] Such evidence of grand correlations failed and global forecasts compromised were gleefully publicized by Richard Proctor, one of Lockyer's most relentless critics.[52]

Proctor reserved equal sarcasm for Baxendell, whose 1876 MLPS paper sought to link rainfall and wind direction to sunspot variations:[53]

> From records of rainfall kept at Oxford it appears that more rain fell under west and southwest winds when sunspots were largest and most numerous than under south and south-east winds, these last being the more rainy winds when sunspots were least in size and fewest in number. This is a somewhat recondite relation, [but] at least proves that earnest search has been made for such cyclic relations as we are considering.[54]

And, as Proctor cheerfully pointed out from other researches by Baxendell, the rainfall-wind relation at St Petersburg was observed to be the precise opposite to that at Oxford.[55] Most damning of all, though, Proctor highlighted sunspot watchers' inability to agree even on the period of sunspot cycles to which correlations should be drawn – these varying in some cases from less than eight years to more than 18.

Stewart's response to such criticisms was to admit that sunspot-weather linkages were of a more complex and multiple character than previously suspected. In 1880, having secured – with unusual diplomacy – data on European rainfall from his successor at Kew (George Whipple), Stewart showed that sunspot-weather correlations were subject not to a single cycle of c. 11 years, but to two distinct periodicities of nine and 12 years.[56] This did not impress Robert Henry

Scott, Stewart's successor as Secretary to the Meteorological Council (formerly Committee). In his 1883 textbook *Elementary meteorology*, Scott contended that 'next to no progress' had been made in the 'cosmical' branch towards understanding the agencies that produced the 'various phases of weather'.[57] Despite Stewart's resolution of complex multiples, Scott devoted an appendix specifically to demolishing the straw target of the 11-year sunspot-weather cycle[58] that he alleged was claimed by the 'high authority' of Meldrum and Stewart. Insofar as the connection was 'not sufficiently understood to justify prediction' and there were 'contradictory conclusions' on the nature of these connections, Scott concluded that it could 'scarcely be said that the close relation between solar and terrestrial phenomena is capable of accurate demonstration'.[59] By 1884, even Stewart's best-informed allies were conceding these points. Professor E. Douglas Archibald, Professor of Mathematics in the Bengal Education Department, who for many years prepared three-day weather forecasts for *The Times* in London, frankly admitted that at present 'we are considerably in the dark about the whole question'.[60]

In 1885 Stewart had retreated (with Schuster's assistance as Langworthy Professor of Physics from 1881) to older speculations on relations between geomagnetism and wind, *viz*. that air currents electrified in the upper atmosphere – perhaps by solar radiation from sunspots – were a major cause of terrestrial magnetic disturbance. Ironically, this was a sideline of his duties as Secretary of a Committee of the British Association for the Advancement of Science set to rework the methodologically-challenged tabulations of the now deceased Sir Edward Sabine.[61] But then Stewart himself died suddenly in late 1887 (only a few months after writing Baxendell's obituary) with the committee still squabbling about the techniques for so doing, a sadly appropriate ending for one unable to secure broad or permanent assent for his views or practices.

After Stewart's demise, and alongside mainstream activities in electrical and X-ray physics,[62] Arthur Schuster built up a major school of meteorology at Manchester that drew strongly on Stewart's prior researches. Schuster adopted a much more limited vision for the field, though, especially now that 'cosmic' meteorology was in decline. He devoted his study of periodicities to the upper atmosphere and to variations on terrestrial magnetism, and was rather more cautious about claims concerning solar influences on the weather.[63] Schuster's caution paid off as he won himself a central place in the meteorological organization of the Royal Society, and thus at the head of the UK's meteorological management – something that neither Baxendell nor Stewart had ever truly accomplished. Schuster's Royal Society obituarist, his protégé G. C. Simpson, argued plausibly that Schuster was responsible for introducing meteorology as a university subject in the UK. In 1905 he set up a small meteorology group within the Physics Department at the University of Manchester and placed Simpson in charge as Britain's first higher education lecturer in the field. This group reorganized the meteorological station in Manchester's Whitworth Park, and, in order to pursue Schuster's agenda of investigating the upper atmosphere, established a kite and balloon station on the hills near Glossop in Derbyshire – where the young Ludwig Wittgenstein experimented in aeronautics until deciding to become a philosopher.[64] From Baxendell's and Stewart's rather controversial and localized nurturance of cosmical meteorology in the preceding century a more 'orthodox' species of terrestrial meteorology spread from Manchester across the country throughout the course of the twentieth century.

Conclusion: gregarious cosmic universalism or comically egregious localism?

For two decades from the early 1860s, Manchester and its Literary and Philosophical Society were the centre of a campaign to launch a cosmical form of meteorology with the fruitful indirect consequences described above. This particular form of civic science, barely hinted at in Kargon's standard account of Manchester's scientific life, reveals particular kinds of intensive networking between meteorologists who were neither simply amateurs nor professionals, nor working within easily recognizable boundaries of disciplinary activity within the physical sciences.[65] Meteorology was to that extent just as much the 'gregarious' science in nineteenth-century north-west Britain as Jim Fleming has shown it to be of nineteenth-century America.[66] The political problems encountered by both Baxendell and Stewart, however, were sufficiently egregious that, in the form they pursued it, it died with them in 1887.

But lest we think of the short-lived and apparently parochial attempt at a universal meteorology as a failed Victorian fad of deluded camaraderie, it is worth noting that just the sort of correlations pursued by Baxendell, Stewart *et al* are today used by a London-based organization called *Weather Action*. In 1994 it used correlations with a *22-year* sunspot cycle to makes private forecasts 'successful' enough to be self-financing (with an annual turnover of £100,000).[67] Although highly controversial still, this organization uses the well-guarded 'Solar Weather Technique' developed by astrophysicist and meteorologist Piers Corbyn to make long-term forecasts that are purchased by insurance companies and other organizations. Such secrecy and commercialization would doubtless have been anathema to the openness and civic altruism of the Mancunian gentlemen discussed above. More than mere anachronistic speculation, such a contrast should remind us of how much has changed in science in the last 150 years, a point that is best understood by regionally specific studies of science such as this. Moreover, bearing in mind the extraordinary developments of interdisciplinary science in the twenty-first century, historians might consider what other uniquely Mancunian activities of Victorian natural science might prove to be similarly revelatory to those seeking to break away from the strait-jacket of twentieth-century conceptions of how knowledge-based specializations emerged.

Notes

1. Part of this paper is drawn from chap. 7, 'Balfour Stewart, exact meteorology and the physical laboratory at Owens College, Manchester', of my doctoral thesis: 'Precision measurement and the genesis of physics teaching laboratories in Victorian Britain' (unpub. PhD, University of Kent at Canterbury, 1989), advisor Crosbie W. Smith. An early version of this paper was presented at the HSS annual meeting in New Orleans, 1994.
2. This activity has often been misleadingly subsumed under the heading 'solar physics': see K. Hufbauer, *Exploring the sun: solar science since Galileo* (Baltimore/London, 1989), p. 49ff.
3. These MLPS publications combined in 1888.
4. The Mancunian roles of Baxendell, Stewart and Schuster are discussed (without reference to mutual meteorological interests and research) in R. Kargon, *Science in Victorian Manchester* (Manchester, 1977), pp. 31, 35, 60, 74–8, 81–4, 195–6, 212–36.
5. 'The Late Mr Joseph Baxendell', *Manchester Guardian*, 10 Aug. 1887, p. 5 col. h.
6. Baxendell's family life was bound up closely with astronomy. As a child he had been introduced to this study by his mother Mary [*née* Shepley] and in 1865 he married Mary Anne Pogson, the sister of Norman Pogson, government astronomer for Madras and fellow observer of variable stars. See J. Bottomley, 'Memoir of the late Joseph Baxendell, F.R.S., F.R.A.S.', *Proceedings of the MLPS*, 4th ser., 1, pp. 28, 30.
7. 'On the variability of ë Tauri', *Monthly Notices of the RAS*, 9 (1846–8), pp. 37–8.
8. J. Baxendell, 'On the variability of 13 Lyrae', *Monthly Notices of the RAS*, 16 (1855–6), p. 201; 'On the period and changes of á Herculis', *Monthly Notices of the RAS*, 16 (1855–6), pp. 201–4; 'On the variable star B.A.C. 3345 (R. Leonis)', *Monthly Notices of the RAS*, 17 (1856–7), p. 235; 'On the variability of 30 Herculis, *Monthly Notices of the RAS*, 17 (1856–7), pp. 266–7.

Notebooks and other documents pertaining to Baxendell's observations on variable stars are held by the Royal Astronomical Society Archives: http://www.ras.org.uk/index.php?option=content&task=view&id=114, accessed 6 Sep. 2005.

9. J. Baxendell, 'On a new variable star (R. Sagittae)', *Proceedings of the MLPS*, 3rd ser., 1 (1857–60), pp. 197–99; 'Observations of the Zodiacal Light', *Proceedings of the MLPS*, 1 (1857–60), pp. 222–5.

10. http://wwp.greenwichmeantime.com/info/railway.htm, accessed 6 Sep. 2005.

11. Baxendell obituary, *Manchester Guardian*, 10 Oct. 1887. On Clare and the history of Quakers in Manchester see http://www.hardshaweastquakers.org.uk/histories/historymanchester.php. For Jones see 'Obituary', *Monthly Notices of the RAS*, 19 (1859), pp. 119–20.

12. A. McConnell, 'FitzRoy, Robert (1805–1865)', *Oxford Dictionary of National Biography* (Oxford, 2004) [http://www.oxforddnb.com/view/article/9639].

13. J. Baxendell, *On the recent suspensions by the Board of Trade of cautionary storm warnings* (privately published, 1867). In the session 1866–7 Baxendell presented three papers on this subject to the MLPS: 'On the recent suspensions by the Board of Trade of cautionary storm warnings'; 'On Dr Buys Ballot's weather signal', and 'On storm warnings'. See Bottomley, 'Memoir', p. 44; and K. Anderson, *Predicting the weather: Victorians and the science of meteorology* (Chicago, 2005), pp. 128, 142.

14. C. W. Sutton, 'Baxendell, Joseph (1815–1887)', rev. Katharine Anderson, *Oxford Dictionary of National Biography* (Oxford, 2004) [http://www.oxforddnb.com/view/article/1727].

15. Later at Southport his advice to take precautionary measures against a smallpox outbreak was overlooked and an epidemic ensued: Bottomley, 'Memoir', p. 32.

16. J. Baxendell, 'On periodic changes in the magnetic condition of the Earth, and in the distribution of temperature on its surface', *Proceedings of the MLPS*, 3 (1864), pp. 251–60.

17. 'Note on Prof Wolf's latest results on solar spots', *Monthly Notes of the RAS*, 21 (1861), pp. 141–3.

18. J. Baxendell, 'On solar radiation, Part I', *Memoirs of the MLPS*, 3rd ser., 4 (1871), pp. 128–39.

19. *Ibid.*, p. 128.

20. *Ibid.*, pp. 130–9.

21. G. V. Vernon, 'Solar radiation observations, made at Old Trafford, Manchester', *Memoirs of the MLPS*, 3rd ser., 4 (1871), pp. 139–42.

22. T. Mackereth, 'A comparison of solar radiation on the grass and at six feet from the ground', *Memoirs of the MLPS*, 3rd ser., 4 (1871), pp. 142–3.

23. T. Mackereth, 'Solar radiation observations, made at Eccles, near Manchester', *Memoirs of the MLPS*, 3rd ser., 4 (1871), pp. 144–7.

24. J. Baxendell 'On solar radiation, Part II', *Memoirs of the MLPS*, 3rd ser., 4 (1871), pp. 147–55, esp. p. 152.
25. B. Stewart, 'Joseph Baxendell, F.R.S.', *Nature*, 36, 1887, p. 585. These passages were muted in the version of the otherwise largely similar obituary that Stewart wrote for the Royal Society: B[alfour] S[tewart], 'Joseph Baxendell', *Proceedings of the Royal Society of London*, 43 (1887), pp. iv–vi.
26. The exception is discussion of his priority disputes with Robert Kirchhoff about the origins of spectrum analysis; see D. Siegel, 'Balfour Stewart and Gustav Robert Kirchhoff: two independent approaches to Kirchhoff's radiation laws', *Isis*, 67 (1976), pp. 565–600.
27. Cited in R. Proctor, *The sun: ruler, fire, light and life of the planetary system* (1871), pp. 207–9.
28. See N. Reingold's entry on Sabine in the *Dictionary of Scientific Biography*; Hufbauer, *Exploring*, pp. 46–51; Proctor, *The sun: ruler of the planetary system* (2nd ed., 1872), p. 195 ff.
29. B. Stewart, 'On sun-spots and their connection with planetary configurations', *Transactions of the Royal Society of Edinburgh*, 23 (1864), pp. 499–504. In the period 1865–73 Stewart published ten empirical papers on this and related subjects in collaboration with Warren de la Rue and Benjamin Loewy: see Royal Society Catalogue.
30. Proctor, *The sun: ruler of the planetary system*, pp. 218–9.
31. B. Stewart and P. G. Tait, 'Preliminary note on the radiation from a revolving disc', *Proceedings of the Royal Society*, 14 (1865), p. 90; idem, 'On the heating of a disk by rapid rotation *in vacuo*', pp. 339–43. Sequels to the lattermost were published in 1867, 1873 and 1878: see G. Gooday 'Sun-spots, weather and the unseen universe: Balfour Stewart's anti-materialist representations of "energy" in British periodicals', in G. Cantor and S. Shuttleworth (eds.), *Science serialized: representations of the sciences in nineteenth-century periodicals* (Cambridge, Mass./London; 2004), pp. 111–47.
32. J. N. Lockyer and B. Stewart, 'The sun as a type of the material universe', *Macmillans Magazine*, 18 (1868), pp. 246–57, 319–27. For further discussion of this essay and other aspect of how Stewart's religious commitments informed his physics, see Gooday, 'Sun-spots, weather and the unseen universe'.
33. *Ibid.*, p. 256.
34. Lockyer and Stewart, 'The sun as a type of the material universe', p. 327.
35. Arthur Schuster, *Biographical fragments* (1932), pp. 208–9.
36. B. Stewart, 'Physical meteorology – its present position', *Nature*, 1 (1869), pp. 101–3, quote from p. 102.
37. Schuster, *Biographical fragments*, pp. 208–9 et seq.
38. B. Stewart, *Recent developments in cosmical physics* (privately published, 1870), copy in Manchester University Library.

39. See Stewart's interviews with the Royal Commission on Scientific Instruction and the Advancement of Science, 1872 and 1874.
40. See Hufbauer, *Exploring*, p. 48ff.
41. See Anderson, *Predicting the weather*.
42. See *First and second reports from the Royal Commission on scientific instruction and the advancement of science, with minutes of evidence, appendices and correspondence, together with a supplementary report and memorial, 1871–72* (Shannon, 1969), pp. 155–6. (The original was published in 1872 by HMSO in London.)
43. R. M. Friedman, *Appropriating the weather: Vilhelm Bjerknes and the construction of modern meteorology* (London, 1989).
44. For example, in 1873 the Secretary of the Mauritius Meteorological Society, Dr Charles Meldrum, reported cyclone-frequency in the Indian Ocean peaked at times of sunspot maxima; likewise the rainfall in Mauritius and Australia: Charles Meldrum, 'On a periodicity of cyclones and rainfall in connexion with the sunspot periodicity', *British Association Report* (1873), pp. 466–78.
45. A. J. Meadows, *Science and controversy: a biography of Sir Norman Lockyer* (1972), pp. 124–9.
46. [Anon.], 'The Unseen Universe', *Nature*, 12 (1875), pp. 41–3; cf. pp. 143–6 of Anon. [Tait and Stewart], *The unseen universe* (1875).
47. *Ibid*. See especially p. 91, also pp. 111–18.
48. See P. J. Hartog, 'Stewart, Balfour (1828–1887)', rev. G. J. N. Gooday, *Oxford Dictionary of National Biography* (Oxford, 2004) [http://www.oxforddnb.com/view/article/26463].
49. As Tait later admitted, the responses to it varied from 'hearty welcome and approval' to 'the extremes of fierce denunciation' or 'lofty scorn': P. G. Tait, 'Dr Balfour Stewart', *Proceedings of the Royal Society*, 41 (1887–8), pp. ix–xi, esp. p. xi.
50. See review by William Clifford, *Fortnightly Review*, 23 (1876), pp. 776–93; and comments by 'E.C.' in *Fraser's Magazine*, 93 (1876), pp. 60–8; [Anon.], 'The Unseen Universe', *Nature*, 12 (1875), p. 43.
51. Meadows, *Science*, p. 127.
52. R. A. Proctor, 'Sunspots and commercial panics', in *Rough ways made smooth: a series of familiar essays on scientific subjects* (London, 1880), pp. 26–31. I am grateful to Bernie Lightman for this reference.
53. Cf. J. Baxendell, 'On changes in the distribution of barometric pressure, temperature, and rainfall under different winds during a solar-spot period', *Memoirs of the MLPS*, 5 (1876), pp. 137–50.
54. Proctor, 'Sunspots and commercial panics', p. 27.
55. *Ibid*. This was almost certainly a slighting reference to J. Baxendall, 'On the distribution of rainfall under different winds, at St Petersburg, during a solar spot period', *Proceedings of the MPLS*, 11 (1872), pp. 135–6.
56. B. Stewart, 'On the long-period inequality in rainfall', *Memoirs of the*

MLPS, 3rd ser., 7 (1882), pp. 161–9.
57. R. H. Scott, *Elementary meteorology* (1883), pp. 1–5.
58. *Ibid.*, Appendix V, 'Note on the relation between sunspots and weather', pp. 392–4; for the increasingly complex character of Stewart's later analysis of multiple periodicities, see A. Schuster, 'Memoir of the late Professor Balfour Stewart', *Memoirs of the MLPS*, 4th ser., 1 (1888), pp. 253–72, esp. p. 264. The mathematics of periodicities later became one of Schuster's specialities.
59. Scott, *Elementary meteorology*, p. 393.
60. E. D. Archibald, 'On the connection between solar phenomena and climatic cycles', in A. Ramsay (ed.), *The scientific roll: a bibliography, guide and index to climate* (1884), pp. 17–37, 145–152; the quote is from p. 152. See Archibald's obituary in *Quarterly Journal of the Meteorological Society*, 40 (1914), pp. 79–80. For some particularly scathing criticisms of meteorology in the early twentieth century, see the discussion of cosmical meteorology by Cleveland Abbe of the US Weather Bureau in 'Meteorology', *Enyclopedia Britannica* (11th ed., 1911).
61. A. Schuster, 'On the connection between sunspots and terrestrial phenomena', *British Association Reports* (1884), pp. 446–63; 'Report[s] of the committee ... appointed for the purpose of considering the best means of comparing and reducing magnetic observations', *British Association Reports* (1885), pp. 65–89; *British Association Reports* (1886), pp. 64–99; *British Association Reports* (1887), pp. 320–35.
62. See P. J. Davies, 'Sir Arthur Schuster, FRS, 1851–1934' (unpub. PhD, University of Manchester Institute of Science and Technology, 1983), which does not discuss Schuster's meteorological activities in detail.
63. A. Schuster, 'On the possible effects of solar magnetization on periodic variations of terrestrial magnetism', *Philosophical Magazine*, 46 (1898), pp. 395–402; 'The application of terrestrial magnetism to the solution of some problems of cosmical physics', *BAAS Report* (1898), pp. 745–7; 'On the investigation of hidden periodicities with application to a supposed 26-day period of meteorological phenomena', *Terrestrial Magnetism*, 3 (1898), pp. 13–41.
64. For six years Schuster even established, at his own expense, a Readership in Meteorology at the University of Cambridge: G. C. Simpson, *Obituary Notices of Fellows of the Royal Society*, 1 (1932–5), pp. 409–23. For Wittgenstein's early career in aeronautical engineering in Manchester see R. Monk, *Ludwig Wittgenstein: the duty of genius* (1990), pp. 28–35; K. Hamilton, 'Some philosophical consequences of Wittgenstein's aeronautical research', *Perspectives on Science*, 9 (2001), pp. 8–10.
65. Kargon, *Science in Victorian Manchester*.
66. J. R. Fleming, *Meteorology in America, 1800–1870* (Baltimore, 1990), preface.
67. P. Coyne, 'Could the sun be the main influence on our weather?', *New*

Statesman and Society, 18 Mar. 1994, p. 47. For information on Weather Action see http://www.weatheraction.com, accessed 6 Sep. 2005. For positive academic appreciation of Weather Action's work see Dennis Wheeler, 'A verification of UK gale forecasts by the "solar weather technique": October 1995–September 1997', *Journal of Atmospheric and Solar-Terrestrial Physics*, 63 (2001), pp. 29–34. See also mention of Piers Corbyn's work in the children's book: John Farndon, *1000 things you should know about Planet Earth* (2000), p. 49.

The early development of scientific research in industry: the case of Metropolitan-Vickers Ltd, 1901–1933

Tim Cooper

Recent years have seen a growing concern with scientific research in industry and its relation to economic growth. Despite such interest, however, professional study of the historical development of scientific research in industry remains rare. For the most part, where histories of individual firms do exist, they are of the celebratory or anniversarial type and even the few written by professional historians rarely pay much attention to research activities.[1] There is certainly nothing in British historiography equivalent to the corpus of American studies on R&D departments in large, high-technology firms, or on biographies of industrial scientists.[2]

There has, of course, been some significant work. The best treatment of the development of British industrial research remains Sanderson's 1972 book *Universities and British industry, 1850–1970*,[3] but, as this covers the whole of Britain over a 120-year period, it does not provide in-depth study at the level of individual firms. While there are some studies which consider the effects of scientific research on particular sectors of the British economy such as steel-making, railways and chemicals,[4] electrical engineering has received little attention, though it is well studied for the US. As Edgerton and Horrocks noted, the character of the scientific research work remains difficult to establish 'in the absence of detailed studies of individual laboratories'.[5] This article aims to make a contribution to this area of study by investigating the development of scientific research at one such company – British Westinghouse, later Metropolitan-Vickers Ltd (known to all in the company as 'MVs').

Getting science into industry

In Britain before 1900, there were few formal research establishments. There was significant technological innovation, not least in the new 'science-based' industries, but it was mainly informal, in workshops and via individual inventions. Contacts between industry and 'pure' science were via the industrial consultancies of university professors, usually for the resolution of specific problems.[6] Formalised industrial research was generally found outside Britain – most notably in Germany for organic chemistry and in the US for electrical engineering. But as Fox and Guagnini note, even these labs were usually chiefly concerned with product and process control and with research to defend patents.[7]

It is difficult to measure the growth of British industrial research (often because of firms' propensity for secrecy), but by 1914 there were at least 15 industrial research laboratories (and in other firms, scientifically-qualified workers, chiefly involved in controlling production, were researching informally on the shop floor).[8] British steel firms from the late-nineteenth century, for example, increasingly used research to fight off competitors and find new markets.[9] Chemical firms, many of which were set up in Britain by German immigrants in the mid- to late-nineteenth century, also made early moves in this direction. They were followed, in electrical engineering, by American firms that set up branches in the UK at the beginning of the twentieth century.

In Manchester several chemical firms had formalised their research well before 1914. The United Alkali Company had a small research department, founded in 1892 with six chemists, and Brunner, Mond used graduate chemists to research the ammonia-soda process. Levinstein's, dye-makers in Blackley, North Manchester (later to become the hub of ICI Dyestuffs), employed a considerable number of graduate chemists for research before the war (perhaps as many as 20). In engineering, Mather & Platt set up their research department in 1908, and the Calico Printers Association established theirs in textile finishing in 1906. The soap manufacturers Crosfields (of Warrington, later part of Unilever) had set up a research department in the 1890s. This is certainly not the complete list: it is likely that organisations such as the Bleacher's Association (cotton finishing), the Co-operative Wholesale Society and Tootal, Broadhurst, Lee (textiles) were employing graduates doing research, even when they did not have formally-constituted research departments.[10]

British Westinghouse came to Manchester in 1901, when the foundation stone of its huge new factory was laid by the Lord Mayors of Manchester and London in the open fields of Trafford Park – England's first industrial park. But initially there was little reason to suppose that the firm would be doing much research itself, or that it would have much of a relationship with the local university, except as a major local employer.[11] The educational institution with which it was most likely to develop a relationship was the local College of Technology, which offered skill training and diploma courses for members of the working class who wished to better themselves.[12]

This being said, the 'holy forty' British engineers who were to manage the firm, and who had gone to Westinghouse's headquarters in Pittsburgh at their own expense in 1899 to spend two years learning how to run a modern electrical engineering concern, returned to Britain with experience of the industrial research departments which were just then being set up in the US.[13] Among them was Arthur Percy Morris Fleming, an English engineer who had graduated from Finsbury Technical College in London and who would eventually set up the works training school and the company's first research department.

A. P. M. Fleming, director of the MVs' research department

Given that British Westinghouse began life as a wholly-owned subsidiary of its American parent, it is difficult to determine when exactly it began to develop its own products. It seems likely, however, that the company's technological knowledge would have come almost entirely from the Westinghouse corporation until 1907, when the latter went briefly into receivership. The British branch was then forced to become much more independent. It was granted permission to sell in markets outside the UK and the dominions, although it struggled in the complex and fractured British market and it was not until 1913 that a dividend on preference shares was paid.[14] Despite the paucity of records for these early years and the lack of a formally-constituted research department, it seems clear that British Westinghouse possessed staff capable of doing research and at least one well-placed staff member keen to set up the kind of R&D operation he had seen at the parent company in the US.

World War One: 'forced' co-operation between industry and academia

The Great War had a huge effect on scientific research, including universities, industrial firms, scientific societies and government provision. After a slow beginning, a spate of industrial 'research' departments were formed in the middle of the war. This was not in order to conduct basic research leading to new inventions, but rather to apply scientific methods to already existing processes, improving testing and quality control and reducing factory costs via rationalisation. Research departments came to centralise these quality concerns in factories which had previously been divided into 'shops' which were run as independent units.

With the universities and industry resolving specific and urgent problems mainly on an *ad hoc* basis, the government began to think about institutional change, though the effects on research and the war effort were limited by their tardy development. The creation of the Admiralty and Army Inventions Boards in 1915, for example, signalled a national concern with military research but also led quickly to interest in civil research.[15] 1916 saw the creation of the Department of Scientific and Industrial Research (DSIR) which, with funding of £1 million, was to aid both the universities and industry. There were grants for research (and thus graduate researchers) of 'special timeliness and promise' and new 'research associations' – bodies funded equally by the government and groups of companies in specific sectors to carry out research of interest to them all. Control of the National Physical Laboratory (NPL) and a number of research boards (such as Fuel and Building) also passed to the DSIR.[16]

In the universities, the start of the conflict resulted in a rapid decimation of student numbers and of younger staff (who often joined up as volunteers), and to calls from professors and scientific institutions for more intelligent use to be made of the expertise which was left.[17] Most of the normal university research and teaching stopped, and staff were called onto official committees to propose methods of applying their scientific expertise to the war effort. University laboratories became central to the mobilisation of science and industry, given the impossibility of setting up new laboratories in government or industrial establishments quickly and at reasonable cost.

The first recognition of the scientific abilities of the nascent British Westinghouse research department appeared in 1915 when the Lancashire Anti-Submarine Committee was formed, comprising scientific and technical experts from all of the research sites in the Manchester region. The composition of this body gives a good idea of the extent of industrial and academic co-operation in the area at the time. The chairman was John Taylor of Mather and Platt, the vice chairman was A. P. M. Fleming at British Westinghouse. The Committee brought together the professors of Physics (Rutherford) and Electrotechnics (Beattie) at Manchester University; the ordinary members included Edward Hopkinson (again of Mather and Platt, and brother of Manchester University's Vice-Chancellor), Dr E. W. Marchant of Liverpool

University and Professor Miles Walker of the Manchester College of Technology (and former research engineer at British Westinghouse).[18] This new grouping thus included not only industrialists and academics, but also a number of experts who had already crossed the line between the two spheres and could understand the views and language of both.

In an example of the fluidity of staff movements between university and government, Gerrard, the electrotechnics lecturer, was sent to the anti-submarine establishment at Hawcraig, Scotland in 1915 and remained working on this subject (which included researches into magnetic mines, hydrophones, valves and the magnetic fields of ships) until the Armistice, when he returned to the department.

Formalising research at British Westinghouse Ltd

Though British Westinghouse did not formally possess a research department before 1916, research in the various 'shops' at their huge factory in Trafford Park was overseen by A. P. M. Fleming. His continued close co-operation with Westinghouse America, and with American practice in industrial research generally, was demonstrated in the middle of the war (1916) when he returned to the US to tour the research departments of a number of companies in order to gain information on how research should be organised in Britain.[19] His trip appears to have had considerable influence in both government and industrial circles, one employee of the company opining that his subsequent report was influential in the creation of the DSIR itself.[20]

This trip also crystallised his plans for a research department at Trafford Park which would be formally separate from the rest of the works. Records from the company archive show that when a Privy Council circular on industrial research in arrived in 1916, requesting information on research efforts in fields of interest to the government, Fleming ordered a summary of the Trafford Park factory's work.[21] Writing to the Works Manager on 19 September 1916, he referred to the fact that a great deal of 'investigational work of a scientific character' was carried on in the various departments in connection with new developments but that the results were 'not properly recorded and often not pushed to a final conclusion'.[22] This fragmented system, he said, led to duplication of work. His arguments in support of a separate research department were as follows:

a) Results would be properly recorded.

b) Facilities available for research would be fully utilized.

c) The facilities of 'the American company' (Westinghouse) in research could better be tapped.

d) The engineering departments would be relieved of the 'hampering' effects of having to do research.

e) It could thus be ascertained whether work had already been done elsewhere; and that an individual trained to do this would be more effective than engineers who were called upon to do such work only occasionally.

f) The experimental work still done in connection with new manufacturing processes might receive guidance and assistance from the central organization, and,

g) Much useful advertising was to be gained from 'technical information known to be of a pioneer character'.[23]

Thus the initial proposal seems to have been more concerned with rationalising a rather fragmented division of factory work than with plunging into new 'research', whether applied or more fundamental. Fleming thought that such a set-up would also be of 'very considerable assistance' to the company's educational work. He had himself been instrumental in setting up in 1910 British Westinghouse's works' training school for apprentices of all levels; they were the first company in Britain to do so.[24]

When formed at the beginning of 1917, the new research section investigated the quality and cost of raw materials, considered how far they could be improved and what substitutions could be made with profit.[25] But before even this effort could begin, what was already done on the shop floor had to be standardised. This meant determining limits of impurity or error and developing testing, inspecting, and analytical purchasing and process specifications for the purpose of 'crystallising in record form the company's experience'. This, to date, had been in the possession of, as Fleming put it, 'individuals whose defection may cause inconvenience'.[26]

By 1917 the chemical department of the research section alone contained 'about 20 workers' and there were separate sections for the analysis of cast iron, steel and ferrous alloys; non-ferrous metals and alloys; gases; oils; fuels and water. Other research sections included a metallurgical and electro-magnetic laboratory. All were concerned to a considerable degree (made more important by wartime conditions) with investigating the quality of incoming consignments of materials. Fleming noted that 'practically the whole of our time and energy has been occupied in building up appropriate means whereby routine tests can be carried out with accuracy and rapidity'.[27]

This improvement in the standards of control led to some considerable early successes. Enamel-covered copper wire (used in winding transformers), for instance, was subjected to technical specification and testing, which led to the rejection of 25 per cent of incoming supplies and forced the manufacturer to improve their product in consequence. The centralisation of supplies and testing of materials also led to considerable savings. Cutting oil, for example, was now mixed on the premises and the net amount saved during 1917 was estimated to be a huge £3,000 (not far off the official budget of the research department as a whole). £500 to £600 had also been saved by recovering metal from solder dross. Fleming estimated the average cost from 1917 of the research department as £24,000 p.a. but noted that 85 per cent of the cost of routine testing was recovered from the works departments for work done on their behalf, though works departments were not required to pay for investigations. This is a crucial detail. MVs' peculiar way of accounting for research meant that only speculative or investigational work was noted in the overall company budget line allotted to the research department. Thus statistics equally quoted for MVs' research largely obscure the effective and amount of work done be the department.[28]

Although the company's work remained concentrated on production and rationalisation, towards the end of the war the new research department also had time to undertake research for, or in conjunction with, outside bodies. Work was done on oils and moulded materials for the Institute of Electrical Engineers (of which Fleming had been a member since the early 1900s). Plans were also being made in 1918 to carry out electrical tests jointly with the Municipal School of Technology – probably via their former colleague Professor Miles Walker. Some experimental work had also been undertaken for the Lancashire Anti-Submarine Committee. The engine department was also consulting the NPL on a number of points – probably via the Professor of Engineering at Manchester University, Petavel, who had been on the company Board since 1911.[29]

As a result of such collaborative work, the question of the appropriate nature of a relationship between university science and industry, had, by the end of the war, become a much-debated topic. In Manchester, the contacts made while working together on specific problems, and the success of such joint efforts, were a boon to those wishing to continue collaboration in the post-war period. These hopes were, however, to be dampened by a number of unexpected factors which appeared immediately after the end of hostilities.

Business and research: the formation of Metropolitan-Vickers Ltd and early research policy

Business reorganisation in the postwar period had a considerable effect on research at British Westinghouse. In 1919, a merger with the arms and engineering concern, Vickers, led to the company being renamed

Metropolitan-Vickers Ltd. Fleming's suggestions for a new research department were greatly affected by the take-over as he had to show how the proposed facilities could be made available for the use of the other companies 'either at present associated or likely to become associated with the British Westinghouse company in this country and abroad'.[30] In this new context, Fleming began to push for the department's development in a direction which was more orientated towards basic, speculative research, using the research department construction going on in other industries to press his case that MVs should not be left behind. As an indication of the importance attached to research by other firms, Fleming quoted three examples: 'The Osram Lamp Company (GEC) have plans prepared for a lab of 45,000 Sq. ft.; Robert Hadfield (steel) is also building one and Messrs Tootal, Broadhurst, Lee and company have extensive plans for education and research and propose to establish a lab devoted solely to textile research'.[31]

MVs' new facilities would, he pointed out, serve in the development of:

1) Experimental work required in the design and testing of apparatus.
2) New processes on a scale sufficiently large to determine the best conditions for manufacture in bulk.
3) New materials and tools.
4) New industries arising from discoveries made in the research department.

It is this last sentence which represents a considerable change from the mere rationalisation of testing and specification which Fleming had been emphasising just two years earlier. To bring about such a new direction, Fleming asked for a building of four floors and an annex; he estimated that the staff required for such a laboratory would be about 100, about half of whom would be women and 25 per cent 'really first class assistants'. The total cost of the new facilities was estimated to be about £30,000, with annual running costs in wages around £20,000 and £10,000 in materials. Fleming thought he would need about three years from the erection of the buildings to bring 'the scheme to maturity'. In view of the post-war shortage of good research workers, Fleming also noted it would be necessary to train 'nearly all' their staff from 'picked university students'. The question of staff, he stressed, was 'all-important and every inducement should be made to secure for the heads of each section the best brains obtainable'.[32] This would, he said, probably meet requirements for five years when further extensions would be needed, partly on account of business expansion, but more likely because he forecast 'increasing dependence' of the works on the research department.

Despite Fleming's pressure for the research department to head into speculative research, he thought that scientific advances impinging on the company's business were at this stage more likely to come from outside the company; for example, from the technical agreement which MVs had made with Westinghouse and the Vickers group companies, or from contact with 'the universities and national labs [...] where very special equipment or experience renders this desirable'.[33] It was to keep track of such developments that a 'technical information' section was created and made responsible for the 'interchange of information with the Research Laboratory at the Pittsburgh [Westinghouse] works as per the technical agreement' which remained even after the Vickers takeover. It was also for the collection of 'new scientific data which might be of use in the industry', linking up with research work of 'other' laboratories.[34]

Thus initially it would appear that Fleming was not entirely sure where useful scientific developments for the company would come from. At this early stage, he considered that the basic research going on in national structures such as the NPL, the newly formed research associations and the industrial research labs of linked American and German companies might all be sources of results. However, it seems equally clear that he was already planning for MVs to have a department sufficiently well equipped to provide some of those results itself.

Fleming continued to link research to education. He knew that it was essential to get the right people for research but that an in-house research facility would also help in the training of other classes of workers. The problem of education was pressing given the effects of war on university entrants and teaching. Whereas in the pre-war days the firm had been able to make a selection from the best men at universities and colleges, many other firms had since 'awakened to the importance of employing scientifically-trained staff' and competition for them after the war was extremely keen. Fleming recommended establishing scholarships to deal with this new competition. There was, he insisted, 'a marked connection between the educational and research work'; the university men the company picked were to be trained partly in the works and partly in the laboratories. This type of training would be of use not only to graduates themselves, but because 'appreciation of the fundamental importance of research can be inculcated through the channels afforded by the works school ... this will prevent young workers adopting that conservative attitude towards industrial development which is often a marked characteristic of adult workers.'[35] To this end, Fleming also proposed common methods for all the associated companies in selecting, educating and training apprentices, suggesting that if this could be achieved then 'the considered opinion of [this] group of companies would carry great weight in influencing other concerns, associations and the government'.[36]

By 1920, perhaps in response to a looming economic depression, considerably more emphasis was placed on the international co-operation side of the research department's work. Contact was kept with America but the take-over by the Vickers group also involved the company in a continental cartel agreement. Apart from the usual reasons Fleming gave for interchange of information (avoidance of duplication, records, etc), he also noted that such contacts had to be kept 'to observe the letter and spirit of the agreements which have been made between the companies in regard to the interchange of technical and manufacturing data'.[37]

Fleming's views of how industrial research worked were also developing. He now viewed the process of industrial research in three stages. These were:

1) The elimination of manufacturing difficulties.

2) The development of new materials and processes and the provision of technical data for design.

3) Pure scientific research having no immediate industrial objective.[38]

The inclusion of point 3 is significant and represented the first explicit move towards fundamental research at MVs. 'At the present [1920]', Fleming said, 'we are dealing almost solely with 1 and 2 because in view of the present high cost of materials and labour [...] we have to be efficient before all else'.[39] Scientific research could only be undertaken 'in connection with work of immediate industrial value'. Along these lines, he noted that 'immediate steps are taken by the Department to apply manufacturing information from Pittsburgh and Baden', though what exactly this consisted of is unclear. What does seem clear is that Fleming found it expedient to point out to the directors of the company that the technical agreement was working and the research department was concentrating only on investigations likely to produce quick results.[40]

The turn towards basic research at MVs

In the following years, Fleming's reports show little concern for the effects of the depression – in fact, they show more and more concern with scientific research. By 1921, Fleming's views had crystallised to such a point that he authored a book together with a colleague, J. G. Pearce, called *Research in industry*, in which they laid out how companies of a certain size should organise their research and build laboratories.[41] Despite the depression, Fleming apparently now felt able to declare openly that the research department should be involved in basic research activities. This was due to one of the research department's first successes, which had resulted from a new direction taken on becoming one of the founders of the BBC (then the British Broadcasting Company).

Fleming used this departure as an example of the utility of the 'pure' research he had advocated, noting that when MVs decided to move into a new field, like radio, the research department allowed them to do so quickly. Radio-receiving apparatus had been, he pointed out, 'an entirely new departure in the company's manufacturing programme and experimental work conducted by the research organisation has been the means of enabling the company to become quickly established in this new field'.[42] It was the research department that conducted preliminary negotiations with the Postmaster, investigating broadcasting in the US and then undertaking broadcasting for the BBC in Manchester. 'The station we operate ranks with the best in existence', said Fleming, and 'an increasing amount of purely scientific investigation is being conducted'.[43]

By 1922, Fleming became even more explicit. The research department had the important function of 'maintaining close contact with every new development, thereby preventing the considerable time lag which often takes place between the discovery of the pure scientists and its commercial application', his first explicit statement of a linear model of scientific development. In regard to education, he was also endeavouring to ensure a good supply of graduates ready to take the research department in this new direction. The works school now had 1,000 trainees with the ratio of applicants to acceptances being ten to one. Unlike GE in America however, 95 per cent of those trained were kept on staff and there were 102 graduate college apprentices (including 20 overseas students). In 1921, the research department had about 130 staff in total.[44]

Keeping up with academia: education and scientific liaison

The school's graduate training quickly bore fruit and led Fleming to another new departure. One of the first recruits to the college apprentice scheme was John Cockcroft who completed his training in 1922.[45] He had joined the company as an apprentice after graduating in electrical engineering at the College of Technology under Miles Walker, and in 1922 he left for Cambridge to join Rutherford while still being funded by MVs. This was apparently at Fleming's suggestion and with a letter of introduction from Walker. That Fleming and Walker could place a man with Rutherford, probably the most famous scientist of his day, spoke volumes about the importance

The MVs' research department, c. 1924

of the contacts both had made on the Lancashire Anti-Submarine Committee and about MVs' reputation in general. Fleming obviously considered this first experiment highly successful. The next year he reported that 'in our research organization special attention has been given, by means of an intelligence service, to the establishment of liaison with the most important sources of scientific information at home and abroad, including the foremost research workers the universities, national and commercial labs and libraries'.[46]

This was the beginning of the 'scientific liaison service' which Fleming ran with his right-hand man, George McKerrow. It is unclear when exactly McKerrow joined the company (1923 or 1924) but he was apparently a nephew of the chairman, Sir Phillip Nash, a linguist who had trained as a scientist in Cambridge and who was equally at home in business and academic scientific circles. Fleming's next liaison candidates were Albert James Bradley, under W. L. Bragg who had replaced Rutherford at the University of Manchester, and T. E. Allibone, who joined the MVs' research department as a staff member from Sheffield University in 1924 and then proceeded to the Cavendish in 1926, again to work under Rutherford. His place at Sheffield was then taken by another liaison PhD student, Charles Sykes.[47]

Clearly, having such men in research positions in university laboratories that might conceivably produce important developments for the electrical engineering industry was a cheap and efficient way of surveying the field – in X-rays (Manchester), metallurgy (Sheffield) and high-tension work and atomic physics (Cavendish). The latter's distance from conceivable commercial application was more than made up for by the prestige attached to having contacts at Cambridge. Through MVs' work on the transformers required for the Cavendish, the research department and the board of the company had open access to college high tables – Fleming, McKerrow and others dining there regularly. The closeness of this relationship was confirmed in 1924 when MVs' new High Tension laboratory, which had begun to do work on dielectrics, was opened at Trafford Park by Rutherford.

Keeping business out of research – the 'formation' of Associated Electrical Industries

In the mid-1920s, a number of important commercial developments had an impact upon the electrical engineering industry. MVs and the Trafford Park labs were affected by British Thomson-Houston's acquisition of a controlling interest in Ferguson Pailin and purchase of Edison Swan, by the passing of the Electricity Supply Act (1926), which resulted in the creation of the national grid, and by the labour unrest which culminated in the General Strike of 1926. In 1927, the merger wave which had swept the chemical industry and led to the creation of ICI, also arrived at the shore of electrical engineering. International General Electric (part of GE America and owners of BTH) took over MVs, and in 1928 Associated Electrical Industries (AEI) was formed. This was a great shock to the proud members of what they, at least, saw as the best industrial research department in the country. As Allibone recounts, 'the shock, I was told by dozens of an older generation, was terrible, to find themselves in bed, so to speak, with their greatest rival, and this shock lasted until all that generation had passed on'.[48]

However, this 'merger' actually had very little effect on MVs, the research efforts of constituent firms remaining almost totally separate. Far more important to MVs' business and research was the decision to build the National Grid. Approximately 30 per cent of all contracts for this work were awarded to MVs and Fleming was, by mid-decade, once again able to claim that his foresight in taking the research department

An early vacuum induction furnace

The X-ray unit

in the direction of high voltages had paid dividends. In very diplomatic language, Fleming noted that 'the policy of the board a few years ago in sanctioning the establishment of high-voltage laboratory facilities has put us in a very strong position and our high voltage laboratories have been in operation continuously in research bearing on the design and investigation of problems connected with the new electricity developments'.[49] He might have added that his approach to education as an integral part of the company's work had also led to avoidance of strikes by the company's workforce in 1926. Such success and the board's support of his department's now proven utility allowed him to commit even more of the research department's resources to basic research.

By 1928, just before the Great Depression, Fleming claimed that 'the developments of the past year, more so than any previous year, have shown the increasing dependence of our industry on new scientific development and the help of scientific knowledge'.[50] The highly competitive conditions in the electrical industry made it increasingly necessary to find new applications for electrical devices which could only, he thought, 'arise through patient and persistent research efforts'.[51] These efforts had indeed borne fruit. Perhaps the key development was C. R. Burch's work on vacuum induction furnaces, when, in attempting to create ever higher vacuums, he discovered a group of oils which had extremely low vapour pressure.[52] Later named Apiezon oils, they were to prove of immense importance for electronic valves (especially the Chain Home radar system built by MVs from 1937), for microscopy and later in the pharmaceutical field.[53] However, perhaps more significantly, the new design of vacuum furnace and its use for the fusion of metal alloys led MVs to direct contact with the Physics and Metallurgy departments at the University of Manchester and its first joint basic research project with the University.[54]

The scientific liaison section instigated just a few years earlier was in full flow by the end of the 1920s. MVs' men placed in university labs were reporting back to the company about work on X-ray examination of crystals which had a bearing on the phenomenon of 'creep' in metals (Bradley had now developed the powder method learned on a sabbatical taken in 1926 in Stockholm). At Sheffield University, Sykes was carrying on Allibone's work on zirconium and other metals for use in alloys while Cockcroft's work ('the study of developments in the atomic theory') was claimed to have a bearing 'on ferro-magnetism and electrical conductivity', and provided a testing ground for MVs' very high voltage equipment. By the end of the 1920s, MVs, under Fleming's guidance, was increasingly moving towards ever more basic research – or co-operative ventures designed to get access to it. In the booming economy of 1928, it must have seemed that the path forward for electrical engineering development was clear. However, the test of whether the new form of industrial scientific development was stable and robust in any economic climate was yet to be made.

No turning back: the Great Depression and its (non-) effect on basic research

The Depression's effect on industrial scientific research in Manchester was remarkable for its mildness. Though some slackening of research work was evident due to economic circumstances, it is significant that by this time research in science-based industry was seen by some firms not only as important enough to merit continued support where possible but as a *way out* of financial difficulty (even in the hardest-hit industries). This, despite the fact that industrial research departments were in financial trouble – though MVs managed to survive the period without laying off graduate staff.[55]

Fleming certainly felt it necessary to defend his section. 'It must be borne in mind', he stated in his report for 1929, 'that researches of a fundamental kind cannot be expected to yield commercial results under, as a rule, four or five years of intensive work, but provision for a period ahead is essential if the company is to be in a position to manufacture, in addition to the very highly competitive lines which represent the bulk of our products today, new and progressive lines of apparatus which [...] for some years are of an almost monopolistic character and yield relatively large profits.' Clearly, he was determined to impress upon the management that research was a long-term activity by nature and examples of earlier research work now bearing fruit were provided to support this claim. He noted that the company's metallurgical investigations in conjunction with Bradley at Manchester University had yielded promising methods of improving steel and the production of certain alloys and that the potential application of low pressures to distillation products (i.e. Burch's work) had led to the establishment of very low pressure apparatus on a commercial basis, as for example, in the company's new departures into the application of the demountable valve. MVs' fundamental work in connection with high-tension phenomena also continued to have an increasing application in the design of high-voltage electrical apparatus.

However, despite such examples, Fleming knew that this appeal to faith in the linear model and insistence on past glories would not be enough. He needed to show results, or at least some concrete evidence of economies made in the department. He was to do this during the depression years in two ways: by appealing to the advantages which had accrued from the relatively cheap scientific liaison system which now saw MVs-funded researchers in a number of university physics departments, and by actually producing some of the 'satisfactory returns' he claimed the department developed as a matter of course.

With regard to the first point, as Fleming noted in the department's 1930 report (which would have been written in spring 1931, the middle of the deepest depression), 'the economic conditions that have prevailed throughout the year have precluded the extension of our research facilities at Trafford Park but to some extent the requirements were met by the extension of our scientific liaison which enables us to take advantage of much outside scientific assistance and gives us access to a great deal of first-class research work being done in various parts of Europe'. But even liaison, by this time, needed to be defended and one of Fleming's main weapons was the threat of missing some key development if it were to be axed. It was essential, he insisted, 'to maintain a close contact [...] to guard against the possibility of a chance discovery of a fundamental nature revolutionizing the whole outlook of the electrical industry'. The rate of scientific progress at the time meant, he insisted, that this risk was one which the organization could not ignore and required that 'every precaution should be taken against surprises of this kind'.[56] But in order to achieve such success, it was necessary to have early and complete information, and to obtain this he insisted on 'very complete and close contact preferably *of a personal nature* between industrial and academic laboratories'.[57] Thus it appears that for MVs, merely following the academic literature in the physical sciences of relevance to the company was not enough. Personal, inside information was also required if these results were to be fully understood.

Despite such assurances, according to Allibone at least, things were serious enough to warrant talk among the board of dividing up the research section during the Depression and this led to Fleming's second response to the threat of funding cuts. In March 1931 he called together all the department's section heads and told them to identify which of their activities could immediately be pushed ahead to develop new products which the company could sell. Such 'non-competitive' (i.e. monopolistic) products were the research section's main hope and its researchers rose to the task. Allibone's department, the high-voltage laboratory, immediately started work on developing for sale a family of X-ray tubes for therapeutic use plus other products such as oscillographs and the first British electron microscope. The idea of having the research sections take the development of instruments to the stage of production was so successful that a New Product Department was later created with the specific task of producing the first few units of such items before production was scaled up and handed on to one of the production departments.[58] An essential element of this new departure was the scientific liaison with cutting-edge university labs which allowed Trafford Park staff to identify and develop the commercial possibilities of laboratory instruments used in their university researches.

On the back of this success and from the healthier climate of 1932, Fleming summed up his view of how research departments should deal with economic crises:

> The attitude of industry towards research has undergone very considerable changes in the past ten years. Originally the pursuit of research in industry was confined to the wealthy corporations who sought to maintain an enduring monopoly in their particular field […]. The next stage was a more enlightened attitude on the part of the manufacturer and culminated in this country in the support of Research Associations and the setting up of a number of research Laboratories. […] There was a definite feeling that research was something in the nature of a luxury. Now, however, it has become recognized […] that in times when industrial conditions are unusually severe, research applied to the improvement of new non-competitive products is a most important factor in restoring industry to healthy conditions.[59]

Contact with universities and fundamental work conducted by the department itself now loomed large in the external face of the MVs' research department. Though the number of individuals actually carrying out basic research or co-operating in it was small, the prestige and reputation they procured for the company was considerable. Some of the advances in technique and equipment design made during such co-operation had played a significant part in the research department's survival during the Depression and had gone some way to demonstrating that this new method of organising manufactures was indispensable to a large firm in a science-based industry. The sites in which this work had been carried on were also important. In stark contrast to his statements at the beginning of the 1920s, by the early 1930s the use of and need for technical agreements, government labs and research associations was, Fleming thought, very limited. As he said, 'in view of our technical associations with other organisations it might have been expected that their technical improvements could be absorbed with beneficial results [...]. In the particular activities in which we are engaged, despite the fact that we look with the greatest care and interest to the work of the GE and AEG concerns we can say specifically that we are leading this field in the developments on which we particularly depend'. In fact, apart from some 'occasional inspiration value', the research of the associated companies bore little relevance to the activities in which the Trafford Park Laboratories were engaged. Fleming laid particular emphasis on this point as it was, he thought, 'so contrary to what at first sight might be expected, but which nevertheless long experience proves to be otherwise.'

Conclusion

This investigation of the early development of scientific research shows how MVs' move towards more basic scientific research was, far from being inevitable, gradual and the result of the research director's personal desire to drive industrial research into ever more 'academic' areas. By the end of the 1920s, the activities of the research department of MVs had undergone radical change. The initial conception of the role of a research department in an industrial electrical engineering firm propounded by Fleming in 1916 had, within some ten years, come around to the kind of purpose and direction espoused by similar enterprises in the US. Fleming's scheme for 'scientific liaison' between young industrial researchers and university groups was a cheap way of furthering his aims and may have been unique. Once the company's liaison with academic laboratories became established, the ability of industrial researchers to understand and use academic research results led the company even further down the same path. Economic events changed the nature of the links established with the universities: the speculative research conducted there by 'liaison officers', together with emergency measures taken by Fleming during the Depression, allowed him to claim that, by 1930, the research department and its emphasis on basic research represented a way out of economic difficulty and was a fundamental part of the modern industrial enterprise.

Notes

1. Among such histories of British firms operating in the north west (at the beginning of the twentieth century Britain's, and perhaps the world's,

largest industrial region) are: T. C. Barker, *The glassmakers: Pilkington, the rise of an international company, 1826–1976* (London, 1977); idem, *Pilkington Brothers and the glass industry* (London, 1960); D. C. Coleman, *Courtaulds: an economic and social history* (Oxford, 1980); R. P. T. Davenport-Hines and J. Slinn, *Glaxo: a history to 1962* (Cambridge, 1992); Evans Medical Ltd., *The story of Evans Medical, 1809–1962* (Liverpool, 1962); R. W. Ferrier, *The history of the British Petroleum Company* (Cambridge, 1994); D. W. F. Hardie, *A history of the chemical industry at Widnes I.C.I.* (1950); D. Kennedy, *ICI: the company that changed our lives* (London, 1993); F. D. Miles, *A history of research in the Nobel Division of I.C.I.* (1955); A. E. Musson, *Enterprise in soap and chemicals: Joseph Crosfield & Sons Limited, 1815–1965* (Manchester, 1965); W. J. Reader, *Imperial Chemical Industries: a history* (Oxford, 1970); A. Simon, *The Simon Engineering Group* (Cheadle, 1953); D. Tracey, *Silk and muslin: the story of Tootal Broadhurst Lee & Co., 1816–1963* (1998); B. H. Tripp, *Renold Chains: a history of the company and the rise of the precision chain industry, 1879–1955* (London, 1956); C. Wilson, *The history of Unilever: a study in economic growth and social change* (London, 1954); J. F. Wilson, *Ferranti and the British electrical industry, 1864–1930* (Manchester, 1988). More information can be found in F. Goodall, *A bibliography of British business histories* (Aldershot, 1987).

2. Examples of this approach include: D. A. Hounshell and J. Kenly Smith, *Science and corporate strategy: Du Pont R&D 1902–1980* (Cambridge, 1988); R. R. Kline, *Steinmetz: engineer and socialist* (Baltimore, 1992); L. S. Reich, *The making of American industrial research: science and business at GE and Bell, 1876–1926* (Cambridge, 1985); G. Wise, *Willis R. Whitney: General Electric and the origins of US industrial research* (New York, 1985); M. Graham and B. Pruitt, *R&D for industry: a century of technical innovation at Alcoa* (Cambridge, 1990).

3. M. Sanderson, *Universities and British industry, 1850–1970* (London, 1972).

4. See M. R. Fox, *Dye-makers of Great Britain, 1856–1976: a history of chemists, companies, products and changes* (Manchester 1987); C. Divall, 'Down the American road? Industrial research on the London, Midland and Scottish Railway, 1923–47', http://www.york.ac.uk/inst/irs/irshome/papers/lms.htm; and G. Tweedale, *Sheffield steel and America: a century of commercial and technological interdependence, 1830–1930* (Cambridge, 1987).

5. D. Edgerton and S. M. Horrocks, 'British industrial research and development before 1945', *Economic History Review*, 47 (1994), p. 224.

6. See W. Koenig, 'Science-based industry or industry-based science? Electrical engineering in Germany before World War One', *Technology and Culture*, 37 (1996), pp. 70–101, for a useful discussion of early academia-industry knowledge transfer processes.

7. See R. Fox and A. Guagnini, *Laboratories, workshops and sites: concepts and practices of research in industrial Europe, 1800–1914* (Berkeley, CA; 1999), chap. 4.

8. See Edgerton and Horrocks, 'British industrial research', section II.
9. D. Edgerton, 'British industrial R&D, 1900–1970', *Journal of European Economic History*, 23 (1994), p. 62. See also G. Tweedale, 'Metallurgy and technological change: a case study of Sheffield specialty steel and America, 1830–1930', *Technology and Culture*, 27 (1986), pp. 189–222.
10. Tootals employed their first qualified scientist in 1913 – see Tracey, 'Silk and muslin'. British Westinghouse almost certainly employed graduates on the shop floor from its foundation in 1902: see M. Crane, S. Glow and M. Johnson, *The Metropolitan-Vickers research department* (Worcester, MA; 1990), p. 21. See G. Tweedale and M. Sawai, 'Industrial research in Osaka and the north-west UK from the 1920s to the 1960s', in D. Farnie *et al* (eds.), *Region and strategy in Britain and Japan: business in Lancashire and Kansai 1890–1990* (London, 1999), pp. 252–99, for the most complete list of firms doing research by World War Two.
11. An introduction to the state of scientific research in the university and industry in Manchester in the pre-World War One period can be found in R. Kargon, *Science in Victorian Manchester: enterprise and expertise* (Baltimore, 1977). For an account of the development of Trafford Park, see R. Nicholls, *Trafford Park: the first hundred years* (Chichester, 1996).
12. On technical training in Manchester, see A. Guagnini, 'The fashioning of higher technical education in Britain: the case of Manchester, 1851–1914', in H. F. Gospel (ed.) *Industrial change and technological innovation: a comparative and historical study* (1991), p. 86.
13. Production started at British Westinghouse in 1902, the same year as British Thomson Houston, at GE's subsidiary: See R. J. Jones and O. Marriott, *Anatomy of a merger: a history of G.E.C., A.E.I. and English Electric* (London, 1970), p. 50.
14. *Ibid.*, p. 58.
15. The first bodies of the latter type were the Munitions Invention Department and the Board of Invention and Research – see M. Pattison, 'Scientists, inventors, and the military in Britain, 1915–19: the Munitions Inventions Department', *Social Studies of Science*, 13 (1983), pp. 521–68; and R. and K. MacLeod, 'The social relations of science and technology 1914–39', in C. M. Cipolla (ed.), *The Fontana economic history of Europe*, Vol. 5 (London, 1976), p. 309.
16. See R. S. Edwards, *Co-operative industrial research: a study of the economic aspects of the research associations grant-aided by the DSIR* (1950); and I. Varcoe, 'Co-operative research associations in British industry, 1918–34', *Minerva*, 19 (1981), pp. 433–63.
17. Sanderson, *Universities and British industry*, pp. 217–18. Student numbers at Oxford dropped by 90 per cent, and at UCL and Dundee were halved. In smaller provincial universities some classes – in chemistry at Bangor for example – disappeared altogether. Scientific committees were set up quickly by bodies such as the Royal Society, the Admiralty and the British

Association while professors from all over the country signed a memorandum and held conferences on the 'Neglect of Science' up to 1916.
18. Cf. 'Miles Walker', *Obituary Notices of the Royal Society*, 3 (1939–1941), pp. 779–89. Walker was Professor of Electrical Engineering at the College of Technology, 1912–32.
19. Later published as A. P. M. Fleming, 'Industrial research in the United States of America', in *Technology and Society* (1917).
20. See T. E. Allibone, 'The AEI long term research laboratory: an industrial experiment', *IEE Proceedings*, 134 (1987), pp. 610–13.
21. For an account of the work of the Committee of the Privy Council for Scientific and Industrial Research, see Varcoe, 'Co-operative research', pp. 192–216.
22. Manchester Museum of Science and Industry (hereafter MMSI), Metropolitan-Vickers Archive (hereafter MVA), letter, Fleming to Works Manager, 19 Sep. 1916.
23. *Ibid*. The 'American company' was of course Westinghouse in Pittsburgh with which they had a technical agreement to share information.
24. The school was also unique in that the company trained many more apprentices than it could employ. Its graduates were virtually guaranteed employment and even if they went elsewhere, they retained a strong loyalty to the company. See Crane *et al*, *Metropolitan-Vickers research department*, pp. 25–6.
25. MMSI/MVA Research Department Report (hereafter RDR) 1917.
26. MMSI/MVA. Introduction to RDR 1918.
27. *Ibid*.
28. Thus economic historians' understandable concentration on statistics for such firms can be misleading. In the absence of detailed information on the accounting practices of other large firms, one may question the applicability of their statistical returns.
29. MMSI/MVA. Introduction to RDR 1918. When completed, the company would 'possess a research organisation well ahead of its competitors in this country'.
30. MMSI/MVA. Introduction to RDR 1920.
31. *Ibid*.
32. *Ibid*.
33. *Ibid*.
34. *Ibid*.
35. *Ibid*.
36. *Ibid*.
37. These agreements led to 'intelligence officers' being kept full-time at some of the more important works and engineers from Trafford Park being sent out in this guise to others. This was one of the reasons for which engineers were encouraged to learn foreign languages in the works school.

38. MMSI/MVA. Introduction to RDR 1920.
39. *Ibid.*
40. Though the department continued its (presumably paid) research work for the IEE, BEAMA, the Admiralty (LASC) and the Portsmouth School of Mines.
41. A. P. M. Fleming and J. G. Pearce, *Research in industry* (1922).
42. MMSI/MVA. Introduction to RDR 1921.
43. *Ibid.*
44. Cf. J. Dummelow, *Metropolitan-Vickers 1899–1949: fifty years in brief* (Manchester, 1949).
45. T. E. Allibone, 'Metro Vickers and the Cavendish', in J. Hendry (ed.), *Cambridge physics in the thirties* (Bristol, 1984).
46. MMSI/MVA. Introduction to RDR 1922.
47. Five other research department staff plus the four mentioned above went on to become Fellows of the Royal Society – an achievement for an industrial research lab which Allibone rightly judges to be unique.
48. Allibone, 'The AEI Long Term Research Laboratory', p. 611.
49. MMSI/MVA. Introduction to RDR 1928.
50. *Ibid.*
51. *Ibid.*
52. See Allibone, 'Metro Vickers and the Cavendish' for a more detailed account.
53. Allibone states that patents taken out on these oils were worth over a million pounds to the company and expresses his consternation that Burch did not receive a Nobel prize for this work.
54. See T. Cooper, 'The science-industry relationship: the case of electrical engineering in Manchester 1914–1960' (unpub. PhD, University of Manchester, 2003), chap. 3.
55. This represented a considerable change in the attitudes displayed in the post-World War One slump and was in stark contrast to the situation in the US. There, between 1930 and 1933, scientific employment in industrial research fell by 33 per cent: GE laid off 50 per cent of scientific staff and AT&T nearly 40 per cent. Westinghouse stopped all research. See Edgerton and Horrocks, 'British industrial research', n. 77.
56. MMSI/MVA. Introduction to RDR 1930.
57. Fleming noted, '[i]t has been our practice for many years to build up and maintain as strong as possible a bond between our industrial research department and the various existing academic labs both at home and abroad. This has been accomplished so successfully that we now have open access to every academic lab which is at all likely to produce discoveries which will react on the electrical industry directly or not': *ibid.*
58. Allibone, 'The AEI Long Term Research Laboratory', pp. 610–13.
59. MMSI/MVA. Introduction to RDR 1932.

Manchester friends at odds: Michael Polanyi, P. M. S. Blackett and the scientist as political speaker

Mary Jo Nye

In the course of the past twenty years or so, it has become commonplace for historians who write about the social process of scientific knowledge to refer to the work of Michael Polanyi (1891–1976), particularly his books *Personal knowledge* (1958) and *The tacit dimension* (1966). What is widely cited from these books is Polanyi's argument that the roots of scientific *knowledge* lie in scientific *practice*, in other words in a *social* tradition of crafts and skills that is based in a system of apprenticeship. For Polanyi, acculturation into the scientific community, rather than adherence to a scientific method, serves as a demarcation between science and non-science. Scientists constitute their work through what he called a 'republic of science' in which knowledge of the natural world is established.

In writing about the practice of science, Polanyi spoke with authenticity because he was first of all a scientist, a physical chemist of great accomplishment who had headed laboratories in Berlin and Manchester in the 1920s and 1930s.[1] Polanyi's theory of science is embedded strongly in his career experiences in chemistry, especially in Berlin, at the Kaiser Wilhelm Institutes from 1920 to 1933. However, the Kaiser Wilhelm Institutes are not the whole story. If Polanyi spoke of scientific practice with what we may call the authenticity of personal witness, he also spoke with political passion; not simply the passion of a scientist who loved his work, but the passion of an individual who feared that the science which was the polestar 'that had inspired him since his childhood' would not survive the regimes of Nazi Germany, Soviet Russia, and the Cold War political order.[2]

Polanyi's views, then, were a product of the larger political events of the mid-twentieth century as well as of his laboratory life. His role as political speaker was also a response to debates within the British scientific community in which Polanyi found himself after 1933. In that scientific community at Manchester, where Polanyi directed the Physical Chemistry Department, one of his most prominent colleagues was Patrick Blackett (1897–1974), six years Polanyi's junior. Blackett arrived at Manchester in 1937 as head of the Physics Department and a Cavendish-educated expert on cosmic rays and nuclear particles. Blackett became a leader of operational research during the Second World War and a Nobel Laureate in Physics in 1948 for his experimental work with the cloud chamber. Blackett was one of three University of Manchester scientists to receive a Nobel award while teaching in Manchester, along with Ernest Rutherford in Chemistry in 1908, and Archibald V. Hill in Physiology or Medicine in 1922.³

Polanyi and Blackett were not just colleagues. They were very close friends. They, and their wives Magda Polanyi and Costanza Blackett,

P. M. S. Blackett with his wife Costanza and daughter Giovanna. Reproduced with permission of Giovanna Blackett Bloor

regularly saw each other during the Manchester years. Blackett addressed Polanyi by the familiar 'Mischi' and Polanyi signed his letters to Blackett 'Misi'.[4] The Blacketts arranged for the Polanyis to visit them on vacation in Wales.[5] During the war, when the Blacketts lived in London, they stayed with the Polanyis when visiting Manchester.[6] Polanyi and Blackett stayed in touch after Blackett had taken a professorship at Imperial College and Polanyi had become a fellow at Merton College, Oxford. Their talk by then was of their honours and their lecture tours and their experiences of growing older.[7]

All through these years, certainly from the 1930s to the 1950s, Polanyi and Blackett argued. They disagreed on many matters and they sometimes were angry with each other. In a letter in October 1941 Polanyi expressed chagrin at what he thought was a hostile tone in Blackett's conversation with him earlier in the day, noting '[w]e have always disagreed, yet maintained an entirely genuine link of sympathy'.[8] Blackett responded by denying any personal hostility, but admitted hostility to 'some of your views, and as you [have] to mine'.[9] Nor were these controversies confined to the dinner table or staff rooms. For well over thirty years, Polanyi and Blackett participated in heated debates that took place in British newspapers and popular periodicals, on radio programmes of the BBC, as well as in scientific organizations and university and governmental committees. They were scientists who assumed public roles so that their views became widely known.

In the first part of this article, I lay out brief parallel accounts of Polanyi's and Blackett's careers up to the mid-1930s, when their personal lives and political worries intersected. I then focus on political issues that engaged them in the 1930s and early 1940s, noting that the differences between these two friends reflect a major divide among British scientists on matters having to do with national economic organization, the social responsibility of the scientist, and the institutional autonomy of the scientific community. I conclude with analysis of the different tacks Blackett and Polanyi took after the war: Blackett as a controversial public critic of national policies and as an active physicist who served as a university and governmental administrator; and Polanyi as a physical scientist turned philosopher of science, whose writings he intended to be a guide for the future organization of scientific work.

Parallel lives

Michael Polanyi was born in Budapest, the youngest of five children and the son of a civil engineer and railroad entrepreneur who died in 1905. His widowed mother Cecile was at the centre of a group of

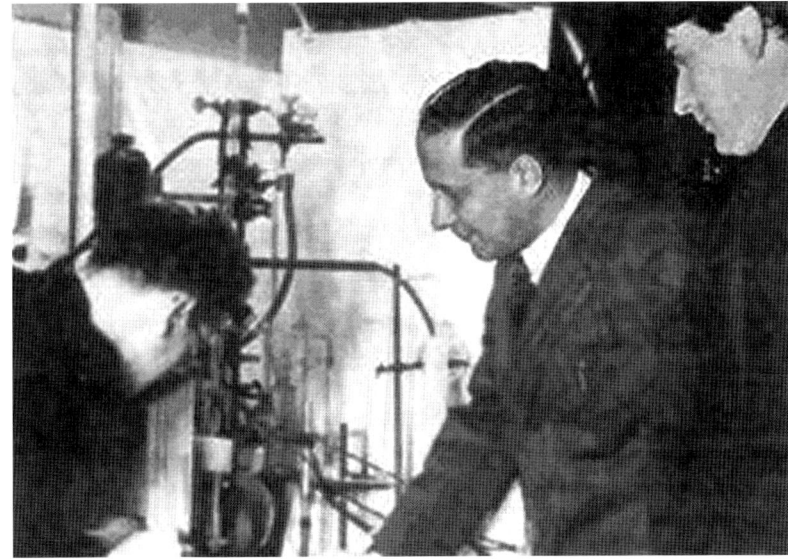

Michael Polanyi in his laboratory. Reproduced with permission of John Polanyi and Istvan Hargittai

intelligentsia that included poets, painters and scholars, most of them sympathetic with bolshevism and socialism. Michael's sister Laura was a feminist and educator who became one of the first women to take a PhD in history in Budapest.[10] His elder brother Karl became an economic historian and published *The great transformation* in 1944, in which he argued the inevitably socially divisive nature of the market economy. In contrast to others in his family, Michael Polanyi, even as a young man, was noted for his scepticism about all forms of socialist ideologies.[11]

Karl and Michael were both active in the Galileo Circle, a Hungarian nationalist group founded in 1908. The two brothers participated in the Sunday Circle, which included the philosopher George Lukacs, the art historian Arnold Hauser, and the sociologist Karl Mannheim.[12] Like many other members of the Budapest intelligentsia, the Polanyi family was originally Jewish, having changed their name from Pollacsek to the more Hungarian-sounding Polányi when they moved to Budapest.[13]

Michael Polanyi completed a medical degree in Budapest in 1913 and entered the Austrian army as a military surgeon, but spent much of the war on leave due to ill health after suffering with diptheria. He began studies in physical chemistry with Georg Bredig and Kasimir Fajans in Karlsruhe in 1913,[14] and in 1917 he finished a doctoral thesis in physical chemistry in Budapest. Polanyi served briefly in the Ministry of Health of what turned out to be the short-lived government of the new Hungarian Republic. In spring 1919, Bela Kun and the Communist Party came to power, followed by a coup d'état at summer's end. The coup's leader was the authoritarian, conservative, and anti-semitic Admiral Nicholas Horthy, who promised a 'Christian' and 'National' government in Hungary.

Although Polanyi was baptized in the Catholic church in 1919, he and others found it increasingly unlikely that scientists of Jewish origin could advance in the university at Budapest during Horthy's regime. Polanyi left Hungary for Germany, first for Karlsruhe and then for Berlin. Later, in a letter to Mannheim, he recalled about his religious views that:

> as a boy and young man I was a materialist and eager disciple of H. G. Wells. My religious interests were awakened by reading *The Brothers Karamozov* in 1913. I was then 22. For the following ten years I was continuously striving for religious understanding and for a time, particularly from 1915 to 1920, I was a converted Christian on the lines of Tolstoy's confession of faith. Towards the middle twenties my religious convictions began to weaken and it was only in the last 10 years that I have returned to them with any degree of conviction.[15]

In Berlin, Polanyi took up a research position at the Kaiser Wilhelm Gesellschaft and married Magda Kemeny, a chemistry student he had met in Karlsruhe.[16] From 1923 to 1933, he was head of a department for reaction kinetics in the Institute for Physical Chemistry and Electrochemistry directed by Fritz Haber. Polanyi's most successful work in this period was the study of fast reaction rates in gases, complemented by his development of a theory of the transition state with the American physical chemist Henry Eyring. He also carried out X-ray diffraction studies of the structure and strength of fibres and

Michael Polanyi with Eugene Wigner and young John Polanyi. Reproduced with permission of John Polanyi and Istvan Hargittai

other solid materials, working with Hermann Mark among others. Polanyi's book *Atomic reactions* appeared in English in 1932.[17]

While Polanyi thought little about religion during the Berlin years, he had not abandoned interest in political matters. In 1928 Polanyi, Leo Szilard, and John von Neumann began following a seminar in economics with Jacob Marschak, trying 'to understand the Russian phenomenon', as Polanyi described it.[18] Russian economics were of more than marginal interest to Polanyi, since his mother was from Vilna, Lithuania and some of his colleagues and relatives were thinking of moving to Russia in positions opening up in the sciences and engineering.[19] Polanyi organized a Sunday circle in Berlin, which met for dining and discussion at Harnack House in Berlin-Dahlem. The group's members included Eugene Wigner, Fritz London, Szilard, von Neumann, and others. Among other things, they debated, like many scientists at the time, whether socialism was better for science and technology than capitalism.[20]

The Berlin days came to an end when Polanyi was dismissed in spring 1933 because of his Jewish origins. Polanyi had just turned down an offer from Manchester in January 1933 and he was able to get it renewed, so that he became head of Physical Chemistry at Manchester. Meredith G. Evans became a close colleague, and Melvin Calvin came from the United States in autumn 1935 to begin a two-year postdoctoral fellowship supported by Polanyi's Rockefeller Foundation funding. Calvin recalled that by the end of his stay, in 1937, Polanyi was losing interest in talking about chemistry and seemed focused on economics and philosophy, which were of less interest to Calvin.[21]

1937 was the year that Patrick Blackett arrived in Manchester as Langworthy Professor of Physics. Blackett was by birth a Londoner, the only son and the second of three children of Arthur Blackett, a stockbroker, and his wife Caroline, daughter of an army officer. Blackett's elder sister Winifred practiced as an architect until she married, and his younger sister Marion became the psychoanalyst Marion Milner. As a 13-year old boy, Blackett entered Osborne Royal Naval College, serving in action throughout the War. After the Admiralty sent Blackett and other young officers to Cambridge in 1919, he resigned from the Navy and embarked upon a physics career, joining Ernest Rutherford's Cavendish Laboratory in 1921.

At the Cavendish, Blackett became one of the pre-eminent experimental physicists of his generation, using the cloud chamber to study high-energy particles. In 1924, from some 23,000 photographs showing 415,000 tracks of ionized particles, he identified eight tracks that showed the path of a proton ejected from a recoiling nitrogen nucleus as an alpha particle was captured by the nucleus, creating an isotope of oxygen. Blackett's classic paper of 1925 included photographs that have been widely reprinted ever since.[22] In the early 1930s, collaboration with Giuseppe Occhialini produced a stunning paper of early 1933 displaying photographs showing the track of a positive electron in a shower of cosmic rays. The experimental success was seated in a clever design using Geiger counters to trigger expansion of gas in the Wilson chamber.[23] Blackett and Occhialini were the first to clearly bring out the idea of pair production as the proper explanation of the experimental effect, making use of Paul Dirac's theoretical prediction of the existence of an anti-electron.[24]

Intersecting political lives

In 1933 Blackett moved to Birkbeck College to head his own laboratory, setting up equipment for detection of cosmic rays 100 feet below ground at an unused platform of Holborn Tube Station. Science journalists wrote stories about Blackett, calling him the new Sherlock Holmes of Baker Street.[25] Just as Polanyi became more politically involved in the mid-1930s, so did Blackett. Whereas Polanyi's formative political milieu had been the radical intelligentsia of Budapest and the mandarin academic milieu of Berlin, Blackett had grown up in London and in the British Navy. As a Bye-fellow at Magdalene College, Cambridge during 1921–3, Blackett became friends with Kingsley Martin, later the editor of the left-wing *New Statesman*. Although Blackett had voted Conservative in 1918, he left Tory politics, campaigning for Hugh Dalton and voting Labour in the 1922 general election. During the General Strike in 1926 he ferried copies of the *British Worker* by car from London to Cambridge.[26] He was active enough in the Labour Party to be asked to run for Parliament representing the University of London in 1935, but he declined.[27]

Blackett later recalled that, for all his interest in science, he had been bored as an undergraduate by standard histories of science and by the way histories treated scientists as if they were 'living so to speak in a social vacuum'. In contrast, he was enthusiastic about Russian papers given at the 1931 meeting in London of the International Congress of the History of Science, where Boris Hessen circulated his famous paper on the 'Social and Economic Roots of Newton's *Principia*'.[28] By 1934 Blackett was speaking on the BBC in a series of programmes on science and society organized by the left-wing scientists Julian Huxley, Hyman Levy and Desmond Bernal. Blackett became one of Gary Werskey's 'visible college' of scientific left-wingers, although not a Marxist.

In BBC talks, Blackett agreed with the argument that science was a social activity, 'a lesson for the scientists to learn, as well as one for statesmen and the lay public,' as Julian Huxley put it.[29] Blackett also agreed with the view that scientists should not delude themselves into thinking that they could be aloof from politics, because their interests were dependent on material and moral support from government and industry. Blackett thought that scientists needed to step up and make clear to the public the virtues and advantages of science at a time when one could see so clearly in Germany in the 1930s the development of anti-scientific and anti-intellectual movements.[30]

However, the view that there could be a scientific consensus in matters of politics was pie in the sky as far as Blackett was concerned:

> No, there I disagree. As a matter of scientific observation, I find that my scientific colleagues, between them, represent all the possible outlooks you have mentioned [...] Don't be too optimistic. I am afraid that if society thinks that the scientist is going to be its saviour, it will find him a broken reed.[31]

On the matter of 'planning,' specifically what was being called 'planned capitalism' in British economic and political circles in the 1930s, Blackett was sceptical. A National government composed of a coalition of Conservatives, Liberals, and some Labour MPs came into power in 1931.[32] Their ideological view was expressed in a manifesto signed in 1935 by a group calling itself 'The Next Five Years' group, including the centrist Tory Harold Macmillan; the Liberal A. D. Lindsay; the editor of *Nature* Richard Gregory; the scientists Julian Huxley, Sir Oliver Lodge and Ernest Rutherford; and the writer H. G. Wells.[33]

'You are being told there is a third way,' Blackett warned, a planned economy as an alternative to socialism or capitalism:

> I believe that there are only two ways to go, and the way we now seem to be starting leads to Fascism; with it comes restriction of output, a lowering of the standard of life of the working classes, and a renunciation of scientific progress. I believe that the only other way is complete Socialism. Socialism will want all the science it can get to produce the greatest possible wealth. Scientists have not perhaps very long to make up their minds on which side they stand.[34]

For his part, living now in Manchester, Polanyi, like Blackett, had objections to planning, but unlike Blackett not to the free-market economy and capitalism. By 1935, Polanyi had visited Russia four times. He afterwards recalled

> a conversation I had with Bukharin in Moscow in 1935. Though he was heading towards his fall and execution three years later, he was

still a leading theoretician of the Communist Party. When I asked him about the pursuit of pure science in Soviet Russia, he said that pure science was a morbid symptom of a class society; under socialism the conception of science pursued for its own sake would disappear, for interests would spontaneously turn to problems of the current Five Year Plan.[35]

Polanyi began to put together a long paper critical of Soviet economics in which he closely examined claims and data from the Soviet Five-Year Plans. The flavour of Polanyi's report can be seen in the following excerpt. In 1933, Polanyi reported, the 'Soviet Union's Red Dawn Knitted Goods Mills received 19 different sets of instructions. The plans were altered as follows: the output plans – 7 times; the productivity of labour plan – 4 times; the cost of production – 8 times. The plan for 1933 was endorsed on January 4, 1934. *This* was central planning', said Polanyi.[36]

Polanyi also began putting together what he called an economic film on 'The Working of Money.' It was a soundless film of diagrams illustrating economic processes with accompanying text. Financial support for improvement of the film came to Polanyi from the Rockefeller Foundation.[37] Selected viewers included members of the Manchester Statistical Society, the technical employees of Imperial Chemical Industries in Northwich, the Hope Street Church Social Study Group in Liverpool, Walter Lippmann's colloquium in Paris, and Friedrich von Hayek's seminar in London.[38] Nor was it only free-market advocates who viewed the film; Desmond Bernal arranged a viewing for the left-wing Association of Scientific Workers in London.[39]

By this time, 1937, Blackett had also moved to Manchester. His acquaintance with Polanyi, begun in Berlin in 1930, became a firm friendship. One of Polanyi's concerns in 1937 was the fate of his niece Eva Striker and her former husband Alex Weissberg, who were in Russian prisons. Ominous word came to Polanyi through scientific friends that he should stop publishing articles against Soviet economics if he did not want harm to come to his family in Russia.[40] Eva was released, but Weissberg remained in prison until he was given over to Nazi troops after the Stalin-Hitler non-aggression pact, announced in August 1939. Blackett wrote to Polanyi from Wales that he would not be surprised if Polanyi felt some *Schadenfreude* over the Russians' behaviour, adding that he himself was completely surprised, as well as dismayed, as Polanyi would expect.[41]

It was clear that there would be war immediately in Europe. Blackett had already been enlisted by Henry Tizard in 1935 to advise the Air Ministry on priorities for national defence. As has been well dis-

Blackett (right) and George P. Thomson

cussed by historians, the Tizard committee argued for development of radar against the objections of the Oxford physicist Frederick A. Lindemann, who was Winston Churchill's scientific adviser. After the Tizard Committee was dissolved, Blackett served on the Maud Committee, constituted in spring 1940, to advise the British government on the feasibility of Britain's developing a fission bomb. Blackett found himself the lone committee member recommending against an independent British effort and urging joint collaboration with the United States, the path that ultimately was followed.[42]

At the outbreak of the war, Blackett joined the instrument section of the Royal Aircraft Establishment where he worked on the design of the Mark 14 bomb-sight, which eliminated the need for a level bombing run at the time of bomb release. In August 1940 he became scientific adviser to the Anti-Aircraft Command, organizing a scientific operational research group to coordinate the use of radar sets, guns and mechanical calculators. He joined the Coastal Command in March 1941, heading a group that recalculated the depth-settings for anti-submarine explosives and vastly improved the use of airborne radar for finding German submarines, and in January 1942 he was transferred to the Admiralty as Director of Naval Operational Research.

In the meantime, Polanyi had asked Blackett to help him get some defence work, but this proved impossible even though Polanyi had become a naturalized British citizen in September 1939.[43] Polanyi instead joined the Oxford zoologist John Baker in founding a scientific association that they called the Society for Freedom in Science. The group first met in March 1941 under the chairmanship of Oxford's retired Professor of Botany, Arthur George Tansley. The group aimed to fight any postwar attempt to allocate funds by central planning in service of social goals on a Soviet-style model.[44] In response to a circular soliciting members, the German refugee physicist Max Born, who was in Edinburgh, wrote to Blackett that he would not join this society since its freedom of science was coupled to an attack on planning and on socialist scientists whom Baker had called 'gangsters' in a review essay attacking Bernal's recent book *The social function of science*.[45]

Polanyi collected a series of his essays against planning, the Soviet Union and Bernalism into a book titled *The contempt of freedom: the Russian experiment and after* (1940), which he sent to Blackett in October of 1941.[46] Blackett found in the work a hostile attitude not toward the Soviet Union alone, but, as Blackett put it, 'to all that is generally called progressive politics – "progressive obsessions" in your words'. The Russians again were allies of the UK following the German army's attack on the Soviet Union in June 1941, and Blackett wrote to Polanyi that he admired what he thought was a remarkable military feat in the Soviet Union's standing up to Germany during the past four months.

During the war, Polanyi, Baker, Tansley, Hayek, the Oxford plant geneticist Frank Darling, the Oxford chemist and historian of science Sherwood Taylor and others argued against government planning for the future of science, linking their objections to a rejection of economic interpretations of the history of science.[47] As news began to circulate about the arrest in the Soviet Union of the geneticist Nikolai Vavilov and the rejection of modern genetics, the virtues of Soviet science and the Soviet system became harder to defend in the UK[48] and anti-Soviet and anti-planning rhetoric became increasingly strident. Archibald Hill, now secretary of the Royal Society, had some sympathy with the Freedom group, but he cautioned Tansley to moderate the group's adversarial language, saying:

> Remember, that Haldane and Blackett, for all their queer political notions, are useful and cooperative members of the [Royal Society] Council: I am sure that Bernal and Hogben will be the same when their turn comes to serve … We can keep them in order better by cooperating with them in scientific affairs than by formally setting up to oppose their political ideas in the name of science.[49]

By 1945, when the war was over, British scientists from all parts of the political spectrum, whether writing in *Nature* or in the circulars of the Freedom Society, all emphasized the necessity for truth and freedom in science. There really was no argument on this issue. Polanyi was asked to address the Manchester section of the leftist Association of Scientific Workers in May of 1945, and his talk was well received. As for scientists' right to be able to follow research paths of no obvious social utility, Blackett wrote to a colleague that Marxist thinking could go too far in reacting against the science for science's sake attitude. A socialist society should devote an appreciable fraction of its resources to pure science as well as to music and art, he wrote.[50] In fact this had always been his view.[51]

Right after the war there was an immediate effort at planning for British universities. The Royal Society was enlisted to furnish a report on 'The Balanced Development of Science in the United Kingdom' and Blackett was one of the committee's members. Polanyi submitted a document arguing that universities should be left to fill professorships with the most eminent candidates available, and the needs of industry, medicine, and defence should be considered only as subsidiary factors in the 'relative endowment of various branches of pure science'.[52] In the end, the committee report supported the 'natural' development of science by 'the most distinguished leaders'.[53] Further, under Blackett's presidency, the Association of Scientific Workers, as well as other groups, recommended against the centralization of governmental scientific offices into a single Ministry of Science. The 1950s organization of science in the UK, then, was not much different than in the pre-war period, though the expenditure was much higher.[54]

Taking a stand on atomic weapons and the Cold War

With the war over, Polanyi resigned his Chair in physical chemistry in 1948 and moved into a professorship created for him at Manchester in 'social studies'. Blackett returned from London to Manchester where he became Dean of the Faculty of Science, and later, Pro-Vice Chancellor, while directing the physical laboratories, including cosmic-ray research, and immersed himself personally in new research on the origins of the earth's magnetism.[55]

Blackett also simultaneously launched a radical public campaign against postwar atomic weapons policy in the United States and the UK, arguing in favour of a neutralist foreign policy for the UK. His widely-read book *Military and political consequences of atomic energy* appeared in 1948, with an American edition published in 1949 under the title *Fear, war, and the bomb*. Blackett republished a series of his essays written over 20 years in 1962 in the book *Studies of war*. In all this, Blackett argued for maintaining contact and cooperation with the Soviet Union. Blackett's attitude not only continued to be one of hopefulness about the Soviet Union, despite the falling of the Iron Curtain over Eastern Europe, but an attitude of distrust toward the United States. Blackett attacked the whole strategic argument for winning wars from the air, with or without atomic weapons, reviving objections that he had made from within wartime operational research against saturation bombing of civilian housing in German cities. This strategy, Blackett wrote in 1948, constituted 'a technically debased form of the original conception' of bombing military and industrial targets, and the debased strategy had led to present acceptance of tactics of mass destruction as a normal operation of war. This strategy, Blackett argued, was ethically immoral and, on the basis of published military reports on the bombing of Germany and Japan, militarily ineffective.[56]

The postwar McMahon Act, which forbade any sharing of American atomic information with another foreign power, even with the UK, was a grave mistake in Blackett's view, and one that ignored the legitimate desire of the Soviet Union for the development of new sources of power from atomic energy. Most controversially of all, Blackett expressed extreme worry that the United States was not open to meaningful negotiations or accommodations with the Soviet Union. Indeed, Blackett argued, and he was the first to make this argument publicly, that the United States had used atomic weapons in Japan in early August 1945 'not so much as the last military act of the Second World War, as the first act of the cold diplomatic war with Russia' to ensure that the Soviet army would not invade Manchuria.[57]

Blackett's arguments were highly controversial and heatedly debated, resulting in *ad hominem* attacks on his integrity and expertise both from within and without scientific communities.[58] Many of his Navy colleagues supported his views. Blackett never reneged on arguments that conventional weapons would not become outmoded because of atomic weapons, nor on his view that conventional and nuclear disarmament should be negotiated in parallel.[59] Nor did he doubt his expertise in this discussion. To the comments of one reviewer that '[s]cience and politics do not readily mix together – at least not in one person',[60] and another that 'because a man is a success in physics, it does not follow that he is well qualified to elucidate political issues',[61] Blackett's rejoinder was obvious. 'Why I should stick to Physics … I cannot quite conceive. Anyway I have spent eleven years of my life in warfare. That gives me a title to talk about it.'[62]

What were Polanyi's views on these matters? He might have adopted the stance of his friend Edward Shils, who characterized Blackett's 1948 book as Stalinist apologia.[63] However, Polanyi in fact was more favourable to Blackett's concerns. In 1946 Polanyi had returned from a trip to the United States during the autumn election

News Chronicle,
7 Mar. 1950

campaign 'shaken to the core', as Blackett wrote to his friend Nevill Mott, by what Polanyi perceived as an almost cavalier attitude in the American press towards the use of atomic weapons.[64] In March 1947 Polanyi wrote a memorandum supporting renunciation of the use of weapons of mass destruction, both atomic and biological, and advocating the adoption of international cooperation for developing atomic energy. Polanyi's view was that the United States should destroy its existing atomic bombs. Polanyi's son, John Polanyi, would later become a leading figure in the Pugwash movement.[65] Blackett, although not Michael Polanyi, began participating in Pugwash conferences after the organization's founding in Nova Scotia in 1957, as did some of his British colleagues including Edward Bullard, Cecil Powell, John Cockcroft, Nevill Mott and Kathleen Lonsdale.[66]

Two paths, but common bonds

Polanyi devoted himself mainly to writing and lecturing in the 1950s and 1960s, after resigning from the Physical Chemistry Department for the Chair of Social Studies created for him at Manchester. This arrangement suited everyone since Polanyi had requested a two-year leave from heading the Chemistry Department in order to prepare his Gifford Lectures on science and religion for the University of Aberdeen.[67] Polanyi gave these lectures in 1951 and 1952, and they came to form a substantial part of the book *Personal knowledge*.

As he became increasingly thoughtful about spirituality and religion in the late 1930s, Polanyi joined the Moot in 1943, convened by J. H. Oldham, an ecumenical Christian leader. The small group met in a rural setting near Horsham, and its members included Karl Mannheim and T. S. Eliot. Many of Polanyi's ideas on the cultivation and transmission of specialized skills and traditions in scientific life were articulated in this setting, in one instance in specific response to a 1944 presentation by Eliot 'On the Place and Function of Christianity'.[68]

In addition, Polanyi became a member in the early 1950s of the Congress for Cultural Freedom, founded by Melvin Lasky. Polanyi's longtime acquaintance Arthur Koestler delivered the Congress's inaugural 'Freedom Manifesto' to about 15,000 people in Berlin in 1950, with the opening declaration that intellectual freedom is one of the 'inalienable rights of man'.[69] Shils reported in the *Bulletin of Atomic Scientists* on a meeting in Hamburg, noting how conference papers contributed to the idea and ideal of the 'autonomous scientific community' as a social system with its own powers of self-maintenance and self-regulation. Shils especially singled out Polanyi's views in assimilating 'the structure of the scientific world' into the 'free market in the economic sphere.' The conflicting claims for 'planning' and 'laissez-faire' were to be reconciled by letting the community of scientists, not individual scientists, do the planning.[70] The Congress later was embarrassed by public revelation of its clandestine funding by the CIA. Still, a broad political spectrum of 'Free World' intellectuals and artists benefitted from these funds, with George Kennan memorably commenting in 1967 that '[t]his country has no Ministry of Culture, and the CIA was obliged to do what it could to try to fill the gap'.[71]

Polanyi's classic article 'The republic of science' appeared in the first issue of Shils' journal *Minerva*, founded in 1962 as a vehicle for discussion of the themes of learning, science, higher education and government.[72] Polanyi's 'free cooperation of independent scientists […] in a free society'[73] functioned by the same norms formulated independently by Vannevar Bush on the one hand, and Robert

K. Merton on the other, as they all made the case that scientists in an autonomous scientific community are the best authorities to decide the distribution of funds from the central government.[74]

In the meantime, having moved from the University of Manchester to Imperial College, Blackett met regularly in London from 1953 to 1963 with a dining group of scientists, including Bernal and C. P. Snow. They advised Hugh Gaitskell, as Leader of the Labour Party, and then Harold Wilson and his Shadow Minister of Education, Richard H. S. Crossman, on science and technology. The expansion of science and technology became a main platform of the Labour Party in the 1964 General Election, and Blackett served as an advisor in the new Ministry of Technology for several years.[75]

While neither the British nor the American governments enforced planning or eliminated science for science's sake after the Second World War, the issues discussed by Polanyi and Blackett remained controversial ones. Blackett, more than Polanyi, genuinely believed that scientists had a social responsibility to bring their expertise and their work to bear in order to improve citizens' lives. His opposition to United States policy on atomic weapons was based partly in a belief that atomic energy, if developed in cooperation among all scientific nations, could improve living conditions and prosperity not only in the Soviet Union, but in the Third World.[76]

P. M. S. Blackett. Reproduced with permission of Giovanna Blackett Bloor

On the issue of the relative merits of socialism and capitalism as economic systems, Polanyi and Blackett were irretrievably divided, at least in general principles, although Polanyi adopted a Keynsian viewpoint in favour of limited government interventions, just as Blackett rejected planning by centralized authority. It was difficult for the two friends to come to any agreement on the Soviet Union, just as there was little room for agreement between the two on a role for religious sensibility in everyday life. Polanyi's later notion of scientific knowledge as 'personal knowledge' made little sense to Blackett, and was similarly puzzling to scientists who shared Blackett's empiricist and phenomenalist approach to scientific questions. Where Polanyi and Blackett agreed without reservation was in their devotion to experimental science and in their shared belief that scientific investigation has a realist intent. For Blackett, as for Polanyi, science was the polestar of his life.

As scientists who spoke out on political matters and as scientists who argued the relevance of experiences from the scientific life to the larger social fabric, Polanyi and Blackett both took considerable chances and showed considerable courage. Not only did they have different political views, but they also waged politics in different ways, Polanyi in his writings on economics and in his philosophical writings, Blackett through his engagement with atomic energy debates and his active administrative and advisory work in university and government settings. They were both barred from the United States in the 1950s, Blackett because he appeared to be a communist fellow-traveller and Polanyi because he had signed a statement of support for what turned out to be a communist-front organization. What bound Polanyi and Blackett together in their political discussions was the fact that they passionately cared. Their commitment to science, as well as their commitment to politics, bound them together as friends, even as friends at odds.

Notes

1. I am grateful for use of the Michael Polanyi Papers (MPP) in the Special Collections of the Regenstein Library at the University of Chicago; and for access to the Patrick Maynard Stuart Blackett Papers (BP/RS) at the Royal Society, with permission of Blackett's children Nicolas Maynard Blackett and Giovanna Blackett Bloor. The American Institute of Physics Niels Bohr Library provided access to the Sources for History of Quantum Mechanics. Research for my studies of Polanyi and Blackett has received support from the National Science Foundation (grant no. SBR–9321305) and the Oregon State University Thomas Hart and Mary Jones Horning Endowment in the Humanities. I thank Churchill College at Cambridge University, where I have been a Bye-Fellow and a visitor, and the Dibner Institute for History of Science and Technology, where I was a Senior Fellow during 2000–1 when I wrote a first version of this paper. Since my completion of this paper, a new biography of Polanyi has appeared although the authors do not mention the friendship of Polanyi and Blackett: W. T. Scott and M. X. Moleski, *Michael Polanyi: scientist and philosopher* (Oxford, 2005). See R. A. Hodgkin and E. P. Wigner, 'Michael Polanyi, 1891–1976,' *Biographical memoirs of Fellows of the Royal Society*, 23 (1977), pp. 421–48; M. J. Nye, 'Michael Polanyi (1891–1976),' *HYLE: International Journal for the Philosophy of Chemistry*, 8 (2002), pp. 123–7; idem, 'Laboratory practice and the physical chemistry of Michael Polanyi', in F. L. Holmes and T. Levere (eds.), *Instruments and experimentation in the history of chemistry* (Cambridge, 2000), pp. 367–400; W. T. Scott, 'At the

wheel of the world: the life and times of Michael Polanyi', *Tradition and Discovery: The Polanyi Society Periodical*, 25 (1998–9), pp. 10–24. Steve Fuller discusses Polanyi's philosophy of science in his polemical study *Thomas Kuhn: a philosophical history for our times* (Chicago, 2000).

2. From a lecture given in 1944 on the BBC and reprinted in *Science, faith and society* (Oxford, 1946).

3. Michael Polanyi's son John Polanyi, who took his 1946 BSc degree and his 1956 PhD degree at Manchester, received the Nobel Prize in Chemistry in 1986. On Blackett, see M. J. Nye, *Blackett: physics, war, and politics in the twentieth century* (Cambridge, Mass; 2004); and [Sir] B. Lovell, 'Patrick Maynard Stuart Blackett, Baron Blackett of Chelsea,' *Biographical memoirs of Fellows of the Royal Society*, 21 (1975), pp. 1–115; published separately as *P. M. S. Blackett: a biographical memoir* (London, 1976). Also P. Hore (ed.), *Patrick Blackett: sailor, scientist, socialist* (London, 2002.)

4. Magda Kemeny Polanyi was an advanced student in chemistry who stopped short of writing a doctoral dissertation at the Karlsruhe Technische-Hochschule. She enrolled at the Berlin Technische-Hochschule after their marriage in 1921 but did not complete the PhD. Their son John became a chemist and John's brother George an economist.

5. Handwritten letter to Magda Polanyi from Costanza ('Pat') Blackett, from Penparc, Llanfrothern, Penrhyndeudraeth, Wales, 28 Mar. 1939. Handwritten note from Costanza Blackett ('Pat') to Michael Polanyi ('Misi'), from Penparc, 5 Sep. 1939 (MPP, 3:14).

6. Patrick Blackett to Michael Polanyi [Mischi] from the Admiralty, 28 June 1945, saying that he was arriving by night train Tuesday and looked forward to seeing Polanyi at the committee on Wednesday afternoon 'and to stay with you in the evening. Pat is coming up on Thursday' (MPP, 4:13).

7. Letter from Magda Polanyi to Costanza and Patrick Blackett ('Dear Friends'), from 22 Upland Park Road, Oxford, 13 June 1965, on the occasion of the award of Companion of Honour to Blackett while Michael Polanyi had just returned from the United States (BP/RS: A.67); note from Magda Polanyi to the Blacketts, dated Nov. 1967, on the occasion of Blackett's receiving the Order of Merit while 'Michael is in the U.S.A.' (BP/RS: A.79).

8. Letter from Michael Polanyi to Patrick Blackett, from Department of Chemistry at Manchester, 28 Oct. 1941, signed 'Love to you and Pat, Misi' (BP/RS: J.65).

9. Letter from Patrick Blackett to Michael Polanyi ('Mischi'), from Pitcullen, Middlesex, 3 Nov. 1941. The letter is signed 'With love to Magda, Patrick' (MPP, 4:7).

10. Scott, 'At the wheel of the world', p. 11; and J. Szapor, 'Laura Polanyi, 1882–1957: narrative of a life,' *Polanyiana*, 6:2 (1997), pp. 43–54.

11. Hodgkin and Wigner, 'Michael Polanyi', p. 413.

12. See J. M. Cash, *Guide to the papers of Michael Polanyi. The Joseph Regenstein Library. Department of Special Collections* (Chicago, 1977), pp. 2, 8. See also L. Congdon, *Exile and social thought: Hungarian intellectuals in Germany and Austria, 1919–1933* (Princeton, 1991). Karl Polanyi emigrated to England from Vienna, where he lectured for the Workers' Educational Association, holding classes in buildings in Oxford and the University of London. In 1940 he went to the United States, where he held a professorship at Columbia University after 1947, but lived in Canada because his wife could not obtain a United States visa due to the McCarran Act. See P. Bohannan and G. Dalton, 'Karl Polanyi, 1886–1964,' *American Anthropologist*, 67 (1965), pp. 1508–11.

13. See L. Polányi, 'Cornucopias of history: a memoir of science and the politics of private lives' (reprint, no page numbers), whose grandfather was Adolf Polanyi (reference courtesy of Pnina Abir-Am). She writes that she is one of fewer than ten living individuals born with the name Polanyi and that none lives in Hungary. Also see W. Gulick, Review of Lee Congdon's *Exile and social thought* in *Tradition and Discovery*, 23:2 (1996–7), pp. 44–6, on p. 44.

14. W. T. Scott, 'Michael Polanyi's creativity in chemistry', in R. Aris *et al* (eds.), *Springs of scientific creativity* (Minneapolis, 1983), pp. 279–307, on p. 283; and G. Holton, 'Michael Polanyi and the history of science,' *Tradition and Discovery*, 19:1 (1992–3), pp. 16–30.

15. Quoted in E. J. Nagy, 'After brotherhood's golden age: Karl and Michael Polanyi', in K. McRobbie (ed.), *Humanity, society and commitment: on Karl Polanyi* (Montreal, 1993), pp. 81–112, on p. 87, from letter from Michael Polanyi to Karl Mannheim, in English, 14 Apr. 1944 (MPP, 4:11).

16. Magda K. Polanyi did not forget chemistry. See M. Polanyi, *Technical and trade dictionary of textile terms* (Oxford, 2nd ed., 1967).

17. M. Polanyi and H. Eyring, 'Zur Berechnung der Aktivierungs-wärme,' *Die Naturwissenschaften*, 18 (1930), pp. 914–5, and 'Über einfache Gasreaktionen', *Zeitschrift für physikalische Chemie*, B12 (1931), pp. 279–311; M. Polanyi, *Atomic reactions* (London, 1932). On Polanyi's solid-state and surface chemistry work, see M. J. Nye, 'At the boundaries: Michael Polanyi's work on surfaces and the solid state', in C. Reinhardt (ed.), *Chemical sciences in the twentieth century* (Berlin, 2001), pp. 246–57; and idem, 'Michael Polanyi's theory of surface absorption: how premature?', in E. B. Hook (ed.), *Prematurity and scientific discovery* (Berkeley, 2002), pp. 151–63.

18. Interview of Michael Polanyi with Thomas S. Kuhn, 15 Feb. 1962, pp. 10–11: Sources for the history of quantum physics, Niels Bohr Library, American Institute of Physics.

19. Polányi, 'Cornucopias of history'.

20. Polanyi discussed the meetings in his interview with Thomas S. Kuhn, 15 Feb. 1962, pp. 10–11: Sources for the history of quantum physics, Niels

Bohr Library, American Institute of Physics. Also see letter from Polanyi to Frau Dr Toni Stolper, 25 Jan. 1930 (MPP, 2:6) and handwritten note, undated (MPP, 43:2).

21. M. Calvin, 'Memories of Michael Polanyi in Manchester,' *Tradition and Discovery*, 18:2 (1991–2), pp. 40–2.
22. P. M. S. Blackett, 'The birth of nuclear science', *The Listener* (March 1954), pp. 380–2, one of three BBC talks (BP/RS: B.116); P. M. S. Blackett, 'The ejection of protons from nitrogen nuclei, photographed by the Wilson Method', *Proceedings of the Royal Society*, A107 (1925), pp. 349–60.
23. Interview of P. M. S. Blackett with John L. Heilbron, 17 Dec. 1962, Imperial College, p. 3: Sources for the history of quantum physics, Niels Bohr Library, American Institute of Physics. Of the Geiger counter, Blackett reminisced that at that time, 'in order to make it work you had to spit on the wire on some Friday evening in Lent'.
24. P. M. S. Blackett and G. P. S. Occhialini, 'Some photographs of the tracks of penetrating radiation,' *Proceedings of the Royal Society of London*, A139 (1933), pp. 699–726 (communicated by Lord Rutherford, received 7 Feb. 1933). Also see J. Chadwick, P. M. S. Blackett, and G. Occhialini, 'New evidence for the positive electron,' *Nature*, 131 (1933), p. 473; P. M. S. Blackett, 'The positive electron', *Nature*, 132 (16 Dec. 1933), pp. 917–19.
25. Interview of Otto R. Frisch with Charles Weiner, 3 May 1967, in New York City: Sources for history of quantum physics, Niels Bohr Library, American Institute of Physics, pp. 19–25.
26. Materials from Lady Costanza Blackett that Blackett intended for an autobiography (BP/RS: A.10A).
27. Blackett, 'Interlude on politics', 2 pages, from materials intended for autobiography (BP/RS: A.10A); and letter from Alex Wood to Blackett, dated 5 Feb. 1935 (BP/RS: A.26).
28. Introduction by Blackett to the first J. D. Bernal Lecture, given by Dorothy Hodgkin, 23 Oct. 1969, at Birkbeck College, 8 half-pages, on p. 4 (BP/RS: H.142).
29. J. Huxley, *Scientific research and social needs* (London, 1934), p. x.
30. Blackett, 'The frustration of science,' 13-page typescript based on 1934 BBC talk (BP/RS: H.1).
31. Blackett, 'Pure science: discussion with Professor P. M. S. Blackett', in Huxley, *Scientific research*, pp. 203–24 on p. 224.
32. See *The next five years: an essay in political agreement* (London, 1935). The Macmillan family publishing house also published *Nature*. See G. Werskey, *The visible college: a collective biography of British scientists and socialists of the 1930s* (London, 1988), p. 238, n.
33. See P. Adelman, *British politics in the 1930s and 1940s* (Cambridge, 1987).
34. Blackett, 'The frustration of science,' n. 30.
35. M. Polanyi, *The tacit dimension* (New York, 1966), pp. 3–4.
36. Michael Polanyi, 'U.S.S.R. economics – fundamental data, system and

spirit,' *Manchester School of Economic and Social Studies*, Nov. 1935, pp. 67–89; also published in Polanyi's *The contempt of freedom* (London, 1940).

37. Letter from the Film Centre to Polanyi, 9 May 1938 (MPP, 3:11), followed by Letter from Michael Polanyi to Mr. Sale, 28 Apr. 1939, reporting that he had received £1,000 from the Rockefeller Foundation in order to release the film for general instructional purposes; also Letter from John Jewkes of the Economics Research Section of the University of Manchester to Robert Letort, Rockefeller Foundation in Paris, 18 May 1939, regarding support for continuing experimental work on the production of diagrammatic films illustrative of economic processes (MPP, 3:15).

38. Typescript, 'Memorandum on economic films', 6 pp. (MPP, 3:6); and carbon copy of letter from Michael Polanyi to Charles Vale in London, 4 Sep. 1937 (MPP, 3:9). Hayek, who received the Nobel Prize in Economics in 1974, published *The road to serfdom* in 1944, arguing against government intervention in the free market and suggesting that government manipulation of the economy leads to totalitarianism, as in Germany. In the late 1930s and 1940s Hayek taught at the London School of Economics; he taught at the University of Chicago as Professor of Social and Moral Science, 1950–62.

39. Letter to Polanyi from Association of Scientific Workers, 24 Aug. 1938; and letter to Polanyi from J. D. Bernal, 10 Sep. 1938 (MPP, 3:12).

40. Letter from H. Rabinovitch to Michael Polanyi, from Geneva, 24 June 1937 (MPP, 3:9). Rabinovitch had talked with Frumkin in Moscow. He also was filing an affidavit with the United States consulate on the matter of the emigration of his sister's son from Vienna; and he was corresponding with the German Jewish Aid Committee about getting the release of his sister Sofie's husband, Egon Szecsi, from a concentration camp.

41. Note from Patrick Blackett to Michael Polanyi, from Penparc, 26 Aug. 1939 (MPP, 4:1). On Polanyi's naturalization papers, Costanza Blackett to Michael Polanyi, from Penparc, 5 Sep. 1939 (MPP, 4:1).

42. Carbon copy of letter from Blackett to George P. Thomson, 28 July 1941; and Thomson's reply, 31 July 1941 (BP/RS: J.104). M. Gowing, *Britain and atomic energy, 1939–1945* (London, 1964); and M. Gowing and L. Arnold, *Independence and deterrence: Britain and atomic energy, 1945–1952* (New York, 2 vols, 1972, 1974).

43. Note from Patrick Blackett to Michael Polanyi, from Penparc, 26 Aug. 1939 (MPP, 4:1). On Polanyi's naturalization papers, Costanza Blackett to Michael Polanyi, from Penparc, 5 Sep. 1939 (MPP, 4:1).

44. W. McGucken, *Scientists, society and state: the social relations of science movement in Great Britain, 1931–1947* (Columbus, 1984), p. 272.

45. Letter from Max Born to Blackett, 22 July 1941, from Edinburgh (BP/RS: J.9). Also see McGucken, *Scientists, society and state*, pp. 266–75. Also, J. Baker, 'Counterblast to Bernalism,' *The New Statesman and Nation*, 29 (July 1939), pp. 174–5. M. Polanyi, 'Rights and duties of science,'

Manchester School of Economics and Social Studies (July 1939), pp. 175–93; also published in *The contempt of freedom*.

46. Letter from Michael Polanyi to Patrick Blackett, from Manchester, 28 Oct. 1941 (BP/RS: J:65).

47. For example, Polanyi in *Manchester Guardian*, 7 Nov. 1942; and Polanyi, 'Autonomy of Science', *Memoirs and Proceedings of the Manchester Literary and Philosophical Society*, 85 (1943), pp. 19–38.

48. C. P. Darlington reported to the editor of the *New Statesman* in 1949 that members of the British delegation to the Soviet Union in 1945 were given two accounts of the disappearance of Nikolai Vavilov, the geneticist and brother of the president of the Soviet Academy. They were told that the story meant for internal dissemination was that he had been shot during the war while trying to escape from Russia; the second story, for external consumption, was that he had died at Magadan during the war 'while breeding frost-resistant plants'. In December 1948 H. J. Muller published a report in the *Bulletin of the Atomic Scientists* on the suppression of the Russian geneticists, and Sir Henry Dale resigned from the Soviet Academy of Sciences when its secretary never replied to the request from the Royal Society for an account of the time and place of death of Nikolai Vavilov: in C. P. Darlington, 'Letter to the editor: the Lysenko controversy,' *New Statesman and Nation*, 37 (22 Jan. 1949), pp. 81–2.

49. Quoted from letter from A. V. Hill to Tansley, 6 June 1941, in McGucken, *Scientists, society and state*, p. 288. H. Rose and S. Rose, *Science and society* (Harmondsworth, 1970), pp. 61, 63–64.

50. Quoted in McGucken, *Scientists, society and state*, pp. 350–1, from letter by Blackett, May 1944 (BP/RS: H:9).

51. Blackett, 'Pure science: discussion with Professor P. M. S. Blackett', n. 31; and interview of Brian Connell with Blackett, Anglia TV, 14-page typescript (BP: A.32), pp. 2, 4.

52. Memos A, B, and C from 1945–1946 (MPP, 22:11).

53. Typescript of five pages, undated, but following 11 Mar. 1946 meeting, entitled 'The balanced development of science in the universities of UK' (MPP, 22:11).

54. T. Wilkie, *British science and politics since 1945* (Oxford, 1991), pp. 47–53. Clement Attlee's post-war government favoured Blackett's recommendation for the establishment of a National Research Development Corporation with right of first refusal for patenting results of research supported by public funds.

55. See M. J. Nye, 'Temptations of theory, strategies of evidence: P. M. S. Blackett and the Earth's magnetism, 1947–1952,' *British Journal for the History of Science*, 32 (1999), pp. 69–92.

56. P. M. S. Blackett, *Military and political consequences of atomic energy* (London, 3rd printing, 1948), pp. 9, 14–15.

57. *Ibid.*, p. 120. Gar Alperowitz revived the argument in 1970: 'I think the only

way you can understand why Nagasaki was tripped off, automatically, bing-bing, just like that, with no consideration, is this tremendous rush to end the war–*not* just to end the war before an invasion, but *immediately!* ... What was the rush? Well, P. M. S. Blackett, another Nobel prize winner, saw in 1945 that the only way you could explain that immediate, fast one-two punch, was the fact that the Russians were in fact scheduled to enter the war on August 9'. Quoted in J. R. Ravetz, *Scientific knowledge and its social problems* (Oxford, 1979), p. 64, n. See G. Alperovitz, *Atomic diplomacy: Hiroshima and Potsdam; the use of the atomic bomb and the American confrontation with Soviet power* (New York, 1965); and G. Alperovitz, *The decision to use the atomic bomb and the architecture of an American myth* (London, 1995). W. Kaempffert, 'The atom's power in war and peace: a famous British physicist champions Russia's side in the momentous debates', in the Book Review, Section 7, *New York Times*, 13 Feb. 1949, pp. 1, 32.

58. See Lord Cherwell [F. A. Lindemann], 'Atomic bombing the decisive weapon – and deterrent,' *Daily Telegraph and Morning Post*, 9 Dec. 1948, p. 4; Kaempffert, 'The atom's power in war and peace', pp. 1, 32.
59. P. M. S. Blackett, 'Steps toward disarmament,' *Scientific American*, 206 (Apr. 1962), pp. 45–53, on p. 51.
60. E. M. Friedwald, 'Blackett's book dissected,' *Discovery*, 9 (Nov. 1948), p. 352.
61. Kaempffert, 'The atom's power in war and peace', p. 32.
62. Carbon copy of letter from Blackett to J. Langdon-Davies, n.d. (BP/RS: H.37).
63. E. A. Shils, 'The atomic problem: Professor Blackett's book,' *Manchester Guardian*, 5 Nov. 1948, p. 4. Shils's angry review of Blackett's book appeared in the same issue of the *Manchester Guardian* as the announcement of Blackett's Nobel Prize in Physics (on p. 5). Also, Shils, 'Blackett's apologia for the Soviet position', *Bulletin of Atomic Scientists*, 5 (Feb. 1949), pp. 34–7. At this time Shils was a Reader in Sociology at the London School of Economics and an Associate Professor in the Committee on Social Thought at the University of Chicago. Shils met Polanyi through Leo Szilard, and Polanyi became a regular visitor to Chicago in the next decade with Shils making every effort to get him a permanent appointment at the University of Chicago. See S. Turner, 'Edward Shils, 1910–1995,' *Tradition and Discovery*, 22:2 (1995–6), pp. 5–9, on p. 6; and E. Shils, 'Thirty years of *Minerva*', *Minerva*, Index to Volumes I–XXX, 1962–92, pp. iii–viii, on p. vii; and Shils, 'Robert Maynard Hutchins,' in *Remembering the University of Chicago* (Chicago, 1991), pp. 191–2.
64. Letter from Blackett to Nevill Mott, 14 Nov. 1946 (BP/RS: D.175).
65. Polanyi, 'Memorandum on the atomic bomb' (written in Holland, Mar. 1947), 7-page typescript with handwritten editing and revisions (MPP, 31:2).
66. J. Rotblat, *Pugwash: the first ten years; history of the conferences of science*

and world affairs (London, 1967). Also see Blackett on Pugwash (BP/RS): J.66–J.71.

67. Scott, 'At the wheel of the world', p. 23.
68. P. Mullins, 'Michael Polanyi and J. H. Oldham, in praise of friendship,' *Appraisal*, 1:4 (Oct. 1997), pp. 179–89, on p. 182.
69. F. S. Saunders, *The cultural Cold War: The CIA and the world of arts and letters* (New York, 1999), p. 82.
70. E. Shils, 'The scientific community: thoughts after Hamburg', *Bulletin of the Atomic Scientists*, 10 (1954), pp. 151–5. See D. Hollinger, *Science, Jews, and secular culture: studies in mid-twentieth century American intellectual history* (Princeton, 1996), pp. 80–96, 155–174; and, more broadly, G. Jones, *Science, politics and the Cold War* (London, 1988).
71. Saunders, *The cultural Cold War*, p. 408.
72. See Turner, 'Edward Shils', p. 5.
73. See Hollinger, *Science, Jews, and secular culture*, pp. 155–74.
74. Hollinger notes how Merton's 1942 article 'A note on science and democracy,' formulating the scientific ethos of universalism, disinterestedness, communism (meaning community) and organized scepticism had become 'The normative structure of science' in Merton's *Sociology of science* (1973). The 1942 article was written originally at the request of Georges Gurvitch, a refugee from France. Also see D. J. Kevles, 'The National Science Foundation and the debate over postwar research policy, 1942–1945: a political interpretation of *Science – The Endless Frontier*,' *Isis*, 68 (1977), pp. 5–26.
75. Lovell, 'Patrick Maynard Stuart Blackett', p. 79.
76. A. M. Weinberg, 'Criteria for scientific choice II: the two cultures,' *Minerva*, 3 (1964), pp. 3–14; and P. Abelson, 'Are the tame cats in charge?', *Saturday Review*, 1 Jan. 1966, pp. 100–3, on p. 102, quoted in J. H. Capshew and K. A. Rader, 'Big science: price to the present', *Osiris*, 2nd ser., 7 (1992), pp. 3–25, on p. 13.

Molluscs, mummies and moon rock: the Manchester Museum and Manchester science

Samuel J. M. M. Alberti

The current collections of the Manchester Museum are based on those of the Manchester Natural History Society and the Manchester Geological Society. They were transferred in 1868 to Owens College, the predecessor of Manchester University, and opened fully in Alfred Waterhouse's grand building in 1890.[1] Like other civic institutions, the Manchester Museum has acted as a node in the network of local scientific practice, and, as a university museum, it had a particularly high profile in this respect. Research activities at the Museum in the century following its opening provide a useful prism through which to examine the mechanics of collaboration within the Museum, the University and the city. In this article, I explore three particular projects, taken from the spectrum of the Museum's activities, to illustrate different aspects of Manchester science at the beginning and end of the twentieth century. They are the morphological work carried out on molluscan collections sent to members of the Museum community at the turn of the century; the studies of unwrapped mummies in 1908 and 1975; and the exhibition of a moon rock from the Apollo 11 mission. I will detail the personnel involved – collectors, curators, scientists – and the networks formed in Manchester by and for the research.

The shifting role of the museum is illustrated by the similarities and differences between these enterprises. The zoological and Egyptological projects were central to the museum's activities at the end of the nineteenth century. Their results lie at the heart of the collection, whereas the lunar exhibition was temporary. The high-profile Apollo 11 mission and the mummy unwrappings have in common a blockbuster appeal, which contrasts with the behind-the-scenes mollusc work. Triangulating these contrasts and parallels will illustrate a number of pertinent issues that are addressed in the closing of this piece: the reasons and rationale for collecting; the roles of amateurs and professionals in the museum sector; the spaces for science in the twentieth century; and the changing role of the museum in culture and scholarship. The projects also reflect the changing nature of

Conus clytospira, Melvill & Standen. From Proceedings of the Zoological Society of London, 1901, plate 21. Manchester Museum Zoology Archive

empire, from the British presence in Egypt and India to the US role in the space race.

Central to the account, however, will be the material culture itself – the role of things. To study object biographies is particularly fruitful in the museum context, as so many museum objects have exotic provenances, from far away and long ago.[2] At a multi-disciplinary museum such as Manchester, they encounter a diverse range of staff and a varied audience. Certain museum objects associated with the research projects enable collaborations and scientific practice. They in turn derive identity from the people with whom they become associated. In this article, I trace the diverse provenances and object biographies of an Indian Ocean shellfish *Conus clytospira*, the preserved Egyptian noble Khnum-Nakht and a tiny segment of lunar material (sample 10085,2). Entwined in the meanings of these objects are the circumstances of their selection and the identities of those who worked on them. They are part of a web of relationships that stretch across the collection, the city and the empire. While exploring their disparate origins, collectors and use, a picture will emerge of the contingent and collaborative character of scientific research in twentieth-century Manchester.

Molluscs

The first object in question began life as the home of a living shellfish somewhere in the Arabian Sea a little over a century ago. Moving along the sea bed, a sandy muddy waste interspersed with occasional sandstone outcrops, the unfortunate invertebrate became stuck to a large cable, was unable to escape the pitch coating, and died.[3] This was the Eastern Telegraph Company cable that at this stage, 125 miles west-south-west of Bombay (now Mumbai), was 45 fathoms

(82 metres) below the surface. On or around 7 September 1899, at the request of Eastern Telegraph Company, the Indo-European department of the colonial government sent the steam ship *Patrick Stewart* to dredge the cable from the sea bed, as it was in need of inspection and repairs.[4] The cable was 30 years old, and a section 13 miles long was to be replaced. On board was First Officer Frederick W. Townsend, the company's chief of telegraph staff (and later to be commander of the *Patrick Stewart*). Townsend, once of Manchester, was an avid student of shellfish, and used his position to collect extensively the molluscan life of the Persian Gulf, the Gulf of Oman and the Arabian Sea. While recovering the old sections of this particular cable, Townsend reaped a bounteous harvest of mollusca, including four or five rare species of the genus *Conus*.

Townsend saw three examples of one distinctive species on the cable on the same day, but the largest, some seven inches in length, slipped from the cable, it being difficult to catch all the material that fell off during inspection. He managed to keep a firm grip on two of the shells, however, including that with which we are concerned – the larger of the pair, 119 mm in length and 37mm at its widest point. Although Townsend was intrigued by this mollusc, he did not keep it. Townsend had a friend, Alexander Abercombie, with whom he shared an interest in collecting the mollusca of Bombay. Abercombie sent some specimens he himself collected to his collaborator, the Manchester-based James Cosmo Melvill. Townsend followed Abercombie's example, and it was to Melvill that Townsend sent the mollusc and its fellow cable victims, as he had all his shells from 1890 onwards.

James Cosmo Melvill (1845–1929) was the son and grandson of senior colonial officials.[5] Educated at Harrow and Trinity College, Cambridge, he settled in Manchester, where he rose to be director of the East India merchant firm of G. and R. Dewhurst. A keen naturalist in his leisure time, he amassed vast collections of molluscs, insects and plants, largely by purchase at auction.[6] Later he sustained and expanded his collections through an extensive web of collectors. Melvill was part of a small group of dedicated conchologists in the 1880s and 1890s who lived and worked around Manchester, alongside the keeper (i.e. director) and two assistant keepers at the Manchester Museum. These were respectively William Evans Hoyle (1855–1926), John Ray Hardy (1844–1921) and Robert Standen (1854–1925). In 1888 they formalized their regular meetings as the Manchester Conchological Society, and seven years later they took over from Leeds as the headquarters of the Conchological Society of Great Britain and Ireland.[7] Some early meeting were held in Melvill's house, but from 1891 the organisation was based at the Manchester Museum. The Museum stored the Society's library and shell collection and in return was supplied with key specimens. Melvill sat on the Museum committee for 15 years, including six as chair, and he was to donate his extensive non-European herbarium to the Museum. He utilised similar contacts and methods to acquire his plant collection as he did his molluscs. The acquisition routes that brought the shell to the Manchester Museum ran parallel to and overlapped with the paths along which plants, rocks and other specimens travelled.

Important as he was, Melvill was only one individual in this complex. The symbiotic relationship between the Conchological Society and the Manchester Museum also involved other society members and Museum staff (often the same people), and Robert Standen in particular. Standen grew up on a Lancashire farm and after teaching for five years in Swinton he worked under the zoologist Arthur Milnes Marshall in the zoology department at Owens College.[8] He published on the county's mollusca, and was the first secretary of the Manchester Conchological Society.[9] Upon Marshall's untimely death, Standen transferred to the Manchester Museum as assistant and secretary, and began work on mollusc collection, having also been elected Honorary Curator of the Conchological Society. In 1896 he was appointed Assistant Keeper for zoology, a post he would keep until his death. Although his interests and responsibilities were wide, he was particularly expert at molluscan identification, and it was this keen eye that led Melvill to seek his help.

With Standen, Melvill undertook extensive taxonomic research on the shells as they came to Manchester.[10] Some arrived in crates of sand rich in tiny species, and the task of sorting and sifting material was onerous (whether by hand or with a sieve). Melvill was able to draw upon the help of a number of amateur members of the Conchological Society, and especially upon Museum staff, including J. Wilfrid Jackson (1880–1978), assistant keeper for geology and Standen's son-in-law. Collecting and identification was a family affair; Melvill collaborated extensively with his own son-in-law, Ernest R. Sykes, who also helped sort the shells. J. Ray Hardy, another assistant keeper, was largely interested in coleoptera, but also took an interest in conchology and helped with the arrangement.

As the leader of this group, Melvill's passion was the identification and naming of novel species, and by the turn of the century he was adept at this.[11] Melvill saw new typological species in the slightest morphological difference – in taxonomic parlance, he was a 'splitter', not a 'lumper'. Seizing on what he perceived to be a new species, he published quickly and fiercely defended his priority. He was undeniably successful, naming over 1,000 new species or varieties, including nearly 600 from the region in which Townsend was collecting.

The shell taken from the cable arrived in Manchester as part of an established system of patronage and exchange. Integral to this system was the classification of species, which imbued these objects with new meaning and status. Because of its inversely conical shell Melvill agreed with Townsend that the shell was part of the genus *Conus*, and Melvill and Standen attributed it to the sub-genus *Cylinder*.[12] The specimen, however, was one which they deemed new to science. They dubbed it *clytospira* – 'illustrious spire'. Thus, as '*Conus clytospira* Melvill & Standen' the shell became indelibly associated with two particular people in its trajectory, with a brief but key moment in its biography. Immediately Melvill and Standen announced their intention to deposit a type (the specimen from which a new species is first described, which becomes the standard for further description) at the British Museum (Natural History); the other was deposited in the Manchester Museum.

Remarkable as the cone may have been, however, it soon transpired that it was not so new. Within a few months, Melvill was dismayed to learn that the French malacologist Félix Pierre Jousseaume (1835–1921) had five years earlier described a smaller example of the same species from the Gulf of Aden, and named it *Conus milneedwardsi* after Henri Milne Edwards (1800–1885), director of the Muséum d'Histoire Naturelle. By the laws of zoological nomenclature so painstakingly established in the previous decades, *milneedwardsi* became the formal name of the species, and *clytospira* the junior synonym. Furthermore, the shell was still later identified as the same species as 'Le Drap d'Or Pyramidal', the Pyramidal Cloth-of-Gold shell, that had been known to some eighteenth-century French collectors.

Nevertheless, the specimen in question would henceforward be associated with Melvill and Standen; the British Museum (Natural History) and Manchester Museum specimens of this species retain the moniker *clytospira* as examples of the name, not the species. The exchange and research that involved *clytospira* was part of a two-decade collaboration between them, which operated alongside a number of other projects with collectors other than Townsend. Along these parallel acquisition routes shells came to them from the Falkland Islands, the Antarctic, Madras, Rhodesia, the Torres Straits and from the Loyalty Islands.[13] The routes then intersected in Manchester, where they were brought together, worked on, selected, and either remained, or continued on their journey to other private collections and museums. Similar and overlapping systems are evident for plants, insects, animals, rocks and fossils, involving amateurs and new professionals across the empire. As part of the Museum's wider collection, the biography of the mollusc would within a few years become linked, via association with Standen, to the second object under scrutiny.

Mummies I

The rock tombs of Deir Rifa (or Rifeh), the cemetery of the ancient town of Shas-hotep, lie six miles south of Asyut in the Nile Valley in central Egypt.[14] Around 1 April 1907, the peace of one tomb was disturbed as a small boy (his name lost to history) squeezed past densely packed grave furniture to find two splendid sarcophagi. He had been sent there after the entrance to the tomb had been uncovered by a workman from Qift known as Erfai, who was employed by the British School of Archaeology in Egypt (BSAE). Erfai had reported his find to the supervisor of the dig, Ernest MacKay (1880–1943), a young field archaeologist from Bristol who assisted in the BSAE digs from 1907 to 1912.[15] MacKay and his team had cleared the descending passage as best they could, but, upon finding the four-foot cavity full of tomb furniture, they sent in the boy. Realising that it was worthwhile, they quickly excavated the tomb, removing everything they found.

The tomb was the grave of two members of the priestly class from the XII dynasty (1991–1783 BC). The hieroglyphs identified them as sons of Aa-Khnumu: one Nakht-Ankh and the other Khnum-Nakht – named for the local deity, the ram-headed god Khnum. (The latter's subsequent history will be followed here.) As well as the sarcophagi, Erfai and the other excavators found canopic jars, exquisite models of Nile boats, and five figurines. Although there has since been considerable debate about the precise nature of their kinship, the set was dubbed 'the tomb of the two brothers'.

The decision as to what would happen to these finds lay with Mackay's mentor, the Honorary Director of the BSAE. He was William Matthew Flinders Petrie (1853–1942), the foremost British Egyptologist.[16] Lacking in formal education, Petrie had been trained by his father as a surveyor before developing a passion for the pyramids. He had been excavating in Egypt since 1880, first independently, then for the Egypt Exploration Fund, and later for the British School of Archaeology in Egypt which he set up from his post as Professor of Egyptology at University College London (UCL). Along with Augustus H. L. F. Pitt-Rivers, Petrie was a pioneer of systematic, scientific methods in archaeological fieldwork. Petrie had supervised excavations at Deir Rifa and Giza in the winter of 1906–7, but had already departed down-country when the tomb was uncovered.

Although the committee of the BSAE was formally charged with the distribution of the masses of material being uncovered at this period, the two brothers effectively passed from MacKay to Petrie, thus becoming associated with the most famous Egyptologist of a generation. Although it was a comprehensive set, once unearthed and studied the tomb contents were most valuable to Petrie in financial terms. After his scholarship and reputation benefited from the excavation, by selling them to a suitable public institution they could be translated into further finds. One such institution that Petrie knew well was the Manchester Museum. Later in 1907, Petrie wrote to the Museum to announce the Deir Rifa tomb group find, and promising them to the Museum, as well as first claim on next season's work, if only £500 might be secured to fund the continuing excavations of the British School. With the Museum committee's approval, the director William Hoyle set about securing donations from the civic great and the good, and within a few weeks had successfully amassed the grand sum of £570 19s.[17]

Perhaps the most significant among those who contributed to the fund was Jesse Haworth (1835–1921), a partner in the Manchester manufacturing firm James Dilworth and Sons. From 1887 until 1892, Petrie's excavations had been carried out with Haworth's backing and that of the London businessman Martyn Kennard (1833–1911), who each received one third of the finds. In 1890, Haworth and Kennard had presented their shares to the newly-opened Manchester Museum. By the early years of the new century, Haworth's share of Petrie's earlier finds were still in the process of being transferred to the Museum, and he and Hoyle were discussing the construction of an extension to the building to house these sizeable collections, so Petrie knew there was timely interest in the north west. He lectured regularly to

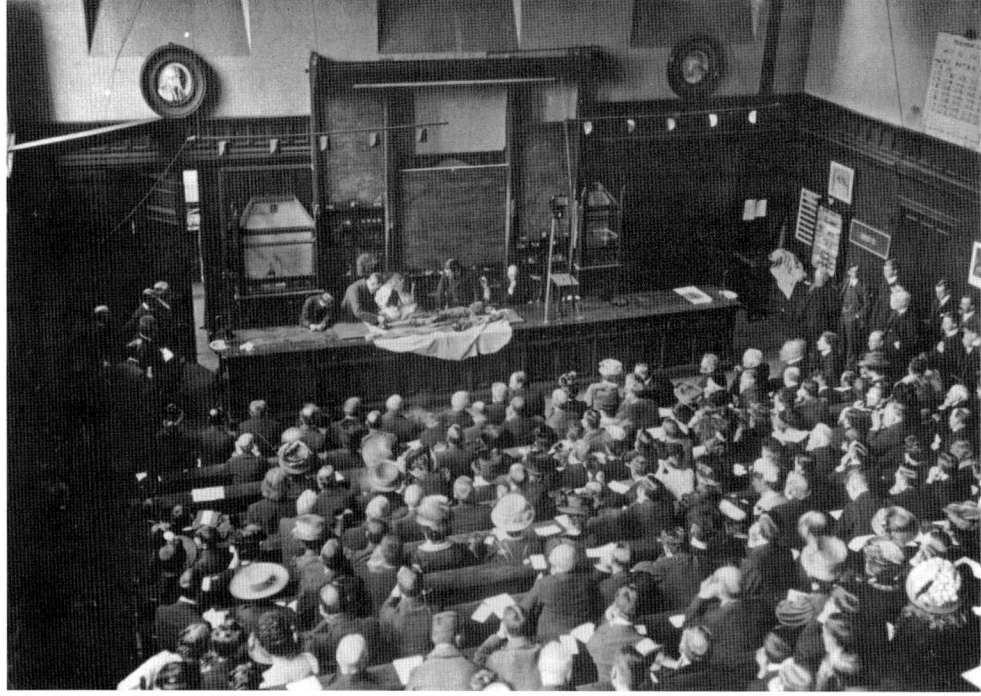

Margaret Murray, Robert Standen and others unwrapping Khnum-Nakht, 1908. Manchester Museum Central Archive

the Manchester Egyptian Society, and was a good friend of Winifred Crompton (1870–1932), at the time the Museum's printer, but later to be appointed assistant in charge of Egyptology.

So it was that the two brothers were sent on the long journey from Deir Rifa to Manchester. They arrived at the Museum in October 1907 and were temporarily exhibited in the Zoology Department.[18] Khnum-Nakht's coffin was absorbed into the collection as item 4725 in the department of Archaeology Accessions register; only later was his body documented as 11725.[19] They joined a collection already strong in Egyptian mummies, but one that did not yet have a dedicated member of staff. But just as the Conchological Society worked in and around the Museum, so too the amateur Manchester Egyptian Association and a loose community of volunteers and visiting scholars kept the Egyptological flag flying. In 1907 they included a visiting lecturer from UCL, Margaret Murray.

Margaret Alice Murray (1863–1963) had intended to follow nursing as a career, but was told she was too short. Instead, she studied under Petrie at UCL and from 1899 was herself a lecturer there. During the academic year 1906–7 she spent time in Manchester giving a course of lectures and cataloguing. And upon the arrival of the two brothers, Murray sought permission to unwrap them. The two brothers themselves were not simply two central objects in this set – as human remains they held a particular status and aura that distinguished them from *Conus clytospira* and from the other objects in the tomb. The corpses were bound up, literally and figuratively, with material culture – the wrappings, the cartonnage, the sarcophagi. In seeking to extract Khnum-Nakht's remains from its physical context, Murray was at once drawing attention to his humanity while emphasizing his value as a specimen. After the removal of the shroud (textile and metaphorical), the body would be incorporated into the collection, assigned a number and rendered a museum object.

Murray's plans for unwrapping were part of a long British tradition.[20] Victorian collectors had carried out unwrappings on mummies arriving in Britain for over a century. Murray's endeavour, however, was intended to contribute not to any parlour-game tradition of 'unrolling', but to the medico-scientific practices of surgeons such as Thomas Pettigrew (1791–1865), who 'unwrapped' in front of esteemed audiences at the Royal Institution, Charing Cross Hospital and the Royal College of Surgeons in the 1830s. Furthermore, the work would be innovative in that the oldest remains hitherto to be unwrapped publicly were from the XVIII dynasty, some three centuries younger than the two brothers.

On Wednesday 6 May 1908 at 2.30pm, Khnum-Nakht was removed from his coffin and placed on the table in the Chemistry Lecture

Theatre (which could hold the largest audience in the University). In the audience were civic grandees and contributors to the purchase fund, members of the Manchester Egyptian Association, invited members of the press, senior Museum staff, and a number of University students. As the *Manchester Courier* reported, 'The learned, the curious, and, it may be added, the uncultured student, gathered in great force in the Chemical Theatre at the Manchester University'.[21] Alas, '[a] number of undergraduates at the back of the chemistry theatre ... were at first inclined to treat the whole affair as a joke'.[22] A correspondent to the *Daily Dispatch* complained, '[i]t is a pity your report did not tell us what the poor, harmless mummy thought about those who had so wantonly disturbed the slumber of the centuries in order that it might be dragged forth to be the butt and the jest of a crowd of students.'[23]

To some, Khnum-Nakht was not an awesome relic but the object of ridicule. To others, he was a scientific investigation. The stage was deliberately quasi-medical: 'Laid on the lecturer's table, covered with a white sheet, was what at first glance might be taken as a "subject" in which a professor of anatomy was to lecture to medical students'.[24] The sheet was unrolled, and Murray began the removal of the ancient textiles. Painstakingly, she peeled off 26 layers of wrapping. Upon removing the final layers, Murray found that the body inside was technically not a mummy; the preservative techniques were not sufficient to embalm effectively. Finally, Khnum-Nakht lay naked and fleshless on the table. What remained of the soft matter was kept for further analysis, and the long process of study began.

As in the case of *clytospira*, Murray did not work alone. Museum staff including Standen, Hardy and Jackson aided in the unwrapping and identifications. Murray also drew on the expertise of the scientific and medical community in Manchester in the analysis of Khnum-Nakht (and of Nakht-Ankh, who was unwrapped quietly later in the year). The situation of the Museum within the University and the academic prestige of the UCL Egyptologists associated with the project drew senior scholars from across the University. Harold B. Dixon, Professor of Chemistry at the University, assessed the salts from the mummy debris. Other colleagues analysed the inorganic constituents, the mummy dust, and the weight, style and colouring of the wrappings.[25] The local physician John Cameron studied the anatomy and pathology of the bodies and the contents of the canopic jars.[26] He was the first to suggest from craniometric measurements that Khnum-Nakht and Nakht-Ankh were so dissimilar that not only were they unlikely to be from the same family, they were not even of the same race, a debate that would continue through the century.

Murray returned to UCL at the end of the academic year. She

slowly worked the results of the study of the analyses into publishable form, with help from Francis Griffith and Petrie. *The tomb of the two brothers* was published by the Manchester Museum in 1910. The two brothers found relative peace on display in the Museum extension that Haworth funded in 1912; Khnum-Nakht's remains remained largely undisturbed for over half a century. But the 1908 unwrapping was not to be the last biomedical Egyptology at the Museum.[27]

Mummies II

In the late 1960s, Theodore Burton-Brown, keeper of archaeology from 1950 to 1969, was approached by Roy White, an amateur Egyptologist who was director of the locally-based commercial X-ray firm, the Radiological Inspection Laboratories (RIL). White wanted to X-ray mummies in the Manchester Museum. Burton-Brown agreed, and his successor John Prag (keeper from 1969) and Donald Ashton (archaeology technician) carried out radiological studies in collaboration with White, occasionally at the RIL base on the other side of the city (via the University's electric mail van), but usually in the Museum with portable equipment. With the help of Roy Garner, appointed zoology technician in 1971, they set the bulky apparatus over the mummies and scanned them.[28]

Garner had come to the Museum from a medical laboratory post in the University, and became interested in the canopic jars containing viscera from Khnum-Nakht and Nakht-Ankh. He took the material to be sampled:

> We ran them through the electron microscope and actually got … surprisingly good images. And it looked as if we might be able to get information out of these mummies. At the time, Rosalie [David] was interested in looking more closely at the mummies, and slowly we began to gather a few people who were interested.[29]

Rosalie David was appointed Assistant Keeper of Archaeology in June 1972, and later became Keeper of Egyptology. She had a degree in Ancient History and Egyptology from UCL and wrote a doctorate in Egyptology at the University of Liverpool, after which she worked for nine months at the Petrie Museum. In Manchester she was given charge of the Egyptian collections at the Museum, their first dedicated Egyptologist for two decades.

David became interested in the scanning techniques being employed on the mummies, and set out to expand this research to establish a multidisciplinary methodology for studying human remains. Approaching Ian Isherwood, Professor of Diagnostic Radiology at the University, she secured space and resources for more detailed

X-ray examination at the Manchester Royal Infirmary. Meanwhile, Garner contacted Richard Pell-Ilderton, a pathologist he knew at the Infirmary, to see if he might be interested in studying the mummies. Pell-Ilderton was coming up to retirement, however, and he recommended a younger colleague, Edmund Tapp of the University Hospital of South Manchester in Withington; Tapp became closely involved for many years. The mummy work soon drew in pharmacists, biochemists, archaeologists, histologists, endoscopists, and even Chief Inspector Tony Fletcher of the Greater Manchester Police, who analysed mummy fingerprints. Many of the other staff of the Museum also became involved at one stage or another. The entomologists Alan Brindle and Colin Johnson, for example, would identify the insects found in the wrappings, as Ray Hardy had seven decades previously.[30]

In the early 1970s there quickly formed a research community surrounding Khnum-Nakht and the Manchester Museum mummies. This community would be at the centre of activities over the following decades, including day schools, two international symposia and later the International Ancient Egyptian Mummy Tissue Bank.[31] In 1974, however, their attention was focussed on one particular event. As the potential for collaboration and skill resources grew, the 'team' decided to repeat Murray's spectacular research and once more to unwrap a mummy. This would be the jewel in the crown of the biomedical research, the nexus around which all the techniques, personnel and press coverage circulated. From the preliminary X-rays David selected mummy 1770, the anonymous and sparsely decorated remains of a young female possibly from the Greco-Roman period from Petrie's excavations at Hawara. The Museum committee agreed to the unwrapping of 1770 in June 1974.[32] Just as Murray had taken Khnum-Nakht to the chemistry department, so a suite of rooms in the Stopford Building of the University's medical school was selected as the site, a few hundred yards south down Oxford Road. On her journey, 1770 passed through the biomedical establishment from museum to hospital to medical school, from store to laboratory.

On 10 June 1975 and over the following fortnight, the unwrapping took place, part-archaeological excavation, part-medical autopsy, in a context not dissimilar to that of 67 years earlier.[33] The Director David Owen sat in the front row of an invited audience of civic and Museum grandees, the mayor and representatives of the Egyptian students' society.[34] Just as in 1908, the print media were in attendance, but this time all stages were photographed and cine-recorded, and BBC cameras were on hand for a planned episode of the *Chronicle* programme.[35] As Murray had before her, David resolved to undertake 'such studies on a scientific and academic basis, instead of the frivolous "unrollings" of mummies performed in front of society audiences in the early nineteenth century.'[36] The mummies were serious, scientific objects. Adorned in medical gowns and masks, David and Tapp laid out medical instruments suitable for an autopsy, including chisels and electric saws. Each layer of bandages and cartonnage was carefully removed, recorded and bagged. Any insects between the layers were set aside for analysis. Funerary decorations on the body were found, including gold finger stalls and nipple covers. Finally the human remains were uncovered, and their position and state recorded.

Khnum-Nakht and his brother were not forgotten, however. 'The two brothers are interesting because they were studied of course by Margaret Murray,' David asserted. '[I]t has been possible to revisit them, and in a couple of instances to redefine earlier conclusions ... they are iconic in that they started Mummy science.'[37] Khnum-Nakht and Nakht-Ankh were redefined, adorned with new layers of meaning. Not only did they provide a bridge between the multidisciplinary unwrapping in 1908 and that of 1975, the remains were used for direct comparison. The dentist Frank Filce Leek (1903–1985), who had been involved in the high profile examination of Tutankhamen in 1968, agreed with John Cameron's findings at the beginning of the century that the cranial and dental morphology of the two brothers were highly dissimilar.[38] Eddie Tapp retrieved the jars in which Murray had stored the remains of tissues from Nakht-Ankh, and undertook rehydration and histological examination; the microscopist Alan Curry applied transmission electron microscopy; and Isherwood concluded from radiological study that Khnum-Nakht did not have a 'club foot'.[39] Khnum-Nakht was no longer the focal point of spectacle, then, but one of a number of pathological specimens. This is not to ignore the continued importance of their visual appeal: the brothers were subject not only to pathological analysis, but also to facial reconstruction.[40]

The brothers' situation as part of a university museum facilitated the research carried out on them. Access to personnel and equipment in the University and teaching hospitals was the strength of the Museum's mummy work. As David reflected, '[t]he unique situation of the Museum with its ready access to University departments and its physical proximity to hospitals with highly specialized, advanced equipment ensured that the project had access to considerable resources'.[41] As well as resources, the expertise that was channelled through the Museum meant that pioneering interdisciplinary work could take place at a time when museums were largely declining as sites for innovative scientific research.

As with all material culture, objects have simultaneously different meanings in different contexts and to different groups. Whether as memento, for aesthetic and cultural appeal, as joke, relic or evidence

Neil Armstrong packing lunar samples on the moon, 21 July 1969. NASA Image AS11–40–5886. National Aeronautics and Space Administration

of palaeopathology, the one object meant different things to different groups as it moved through its acquisition and use. Usually their views were complementary, although some film crews, for whom 1770 was a juicy story, exhibited a different attitude to conservation from Roy Garner and his colleagues. When filming one of the mummies, one camera operator wanted to 'sit on top of the coffin as we pushed the trolley down a hospital corridor so he could get what he called the "mummy's-eye view"', Garner remembered. 'We vetoed that.'[42] As well as objects of research, Khnum-Nakht and the other mummies were among the most news-worthy items in the Museum through the twentieth century, as ancient remains from a land far away. Only one specimen had enjoyed more media attention up to that point. It was very small and it stayed in the Museum for less than a week. But it was far older, and from *much* further away.

Moon rock

At around 04.00 GMT on 21 July 1969, Neil A. Armstrong, civilian commander of the Apollo 11, gathered a small piece of moon 'soil' from the Sea of Tranquillity. These were fine-grained particles of lunar surface, part of the 21.7kg of samples gathered during the first manned mission to the moon.[43] They were gathered with long-handled tongs and scoops and stored in small sample bags. The samples were packed at the modular equipment stowage assembly into a hermetically sealed crate, which was hauled with a lanyard and pulley onto the lunar module, *Eagle*, by Armstrong and his command module pilot Colonel Edwin E. (Buzz) Aldrin. At 17.54 GMT, *Eagle* was fired back into lunar orbit, to dock with the Command Module, *Columbia*, guided by Lt Colonel Michael Collins. The astronauts transferred the samples, and *Eagle* was jettisoned. Three days later, *Columbia* returned to earth, splashing down in the Pacific Ocean.

While Armstrong, Aldrin and Collins were sealed into quarantine conditions for three weeks, the sample crate was unloaded from the capsule onto USS *Hornet* and taken to the Lunar Receiving Laboratory at Houston. There the Preliminary Investigation Team sterilised them with ultraviolet light and germicide and cleaned them with water and nitrogen. On 12 September the National Aeronautics and Space Administration (NASA) began to distribute samples to 300 researchers in laboratories in nine countries for complete analysis, of which the UK would receive 16.[44] Although NASA had originally planned to transport them to the US embassy in London by diplomatic courier, in the event the UK samples were picked up from Houston by the Cambridge petrologist Stuart Agrell (1913–1996) and the astrophysicist Peter Clegg (of Queen Mary College, London).[45] They returned to London on 19 September and took the samples to the Science Research Council headquarters in High Holborn.

That one small sample was to be sent to the University of Manchester was serendipitous. Among those who had applied in 1967 to work on the samples was the prominent mineralogist Jack Zussman, then Reader in Geology at Oxford University. Zussman, after whom Agrell named the mineral 'zussmanite' in 1964, was a crystallographer by training. He had studied at Cambridge and subsequently worked at Manchester for a decade until 1962, when he had accepted the post at Oxford. In 1967 he was appointed to the Chair of Geology back at Manchester, and NASA agreed to allow him to transfer his Principal Investigator status. (Also in Manchester, John Geake at the UMIST Department of Physics was to work with scientists from Hull and Paris on 20g of fine dust to determine surface properties and radiation history of the moon.)

In September 1969, Zussman and the other Principal Investigators set off to Holborn to pick up their samples. During a ceremony that he remembered as 'a little bit like a school prize giving', Zussman was given a container with five grammes of particles and dust of up to a centimetre in size – termed by NASA sample '10085,2' (larger specimens from rock 10044 and further thin sections followed later).[46] With his precious cargo in his pocket, Zussman caught the train home and went directly to the Geology Department, now housed in the Williamson building opposite the Manchester Museum. (Like the other science departments, Geology had moved across Oxford Road during the previous decade, thus ending the direct architectural contiguity between the Museum and laboratory science.) Zussman placed the sample in a combination safe, where it would be returned whenever it was not in use.[47]

The equipment and expertise at Manchester allowed study by optical microscopy, microprobe and X-ray diffraction, as well as electron microscopy and X-ray fluorescence analysis.[48] As Principal Investigator, Zussman collaborated with a team of colleagues in the Geology Department that included William S. Fyfe FRS (b. 1927), Royal Society Professor of Geochemistry in Manchester 1966–72, and William 'Mac' MacKenzie (1920–2001), Professor of Petrology. Despite the tight timetable imposed by NASA, Zussman decided that some of the sample should go on display to the public, to satiate the considerable curiosity, and to promote the work of the University. 'Certainly, everyone was curious to be able to see what moon rock looked like', remembered Zussman. 'We knew that the public were interested, and it would be good to satisfy their interest; good for them and good for public relations for the Museum and for the University to be as open as we could and to show something.'[49]

With its mineralogical and geological displays, the Museum was the obvious venue to advertise this University research. Zussman, MacKenzie and Fyfe, like Cosmo Melvill, all sat on the Manchester Museum committee. Zussman was good friends with the Keeper of Geology at the Manchester Museum, R. Michael C. Eagar (1919–2003).[50] An international expert in fossil non-marine bivalves, Eagar had been at the Museum since 1945 and would remain until his retirement in 1987. Although their professional interests lay at different ends of the earth sciences, Zussman and Eagar had mutual college friends and knew each other through the Senior Common Room. Zussman also worked with Derek Rushton, recently appointed Assistant Keeper under Eagar with responsibility for the rocks and minerals, exchanging specimens for display and research.

Within days of the arrival of the sample, Eagar and Zussman set out to secure permission to display a few small fragments from the soil. With the firm backing of David Owen, Zussman wrote to the SRC, asking if they and NASA would permit a public display. He assured them that security and crowd control would be sufficient during the display, the sample would be returned to the Williamson building each night, and that the exhibition would not detract from the overall investigation.[51] Moon rock from the early samples had been put on display in several sites, including museums, for example the Smithsonian Institution in Washington, and NASA and the SRC gave the go-ahead to Manchester on 26 September. By this time Eagar, Rushton and the Museum's chief technician Harry Spencer had already been frantically at work preparing the exhibition. They planned to display the sample for four days from 30 September, and so had had only a week to prepare what would be the most popular exhibition the Museum ever staged.[52]

The dust was set in a glass dish within a locked case, with a microscope allowing scrutiny. Closed-circuit television cameras magnified the image onto a screen at the entrance to the exhibition. 'Thus, as they entered the exhibition', considered Eagar and Owen, 'people were able to view the material as a scientist might examine it'.[53] Rather than rely on the appearance of the rock itself, they counted on the colossal public interest and the sample's association with Neil Armstrong, now a global celebrity. The Geology Department team were very much involved in setting up the exhibit, despite their pressing research commitment. Dunham supplied photographs. Zussman mounted the dust on glass slides, spent time in the Museum positioning the particles under the cameras, and gave a press conference to announce the exhibition.

Still, the fragments were not stunning objects. Zussman acknowledged,

> If we'd waited we might have had something more spectacular [from later Apollo missions, for example] … [but] we did it early, because the curiosity was there. […] People filed in and looked at it, and they went away I think feeling that they were glad they came and saw it, so they could say, 'I've seen a bit of the moon', but I don't think it was terribly photogenic … I heard one lady say, "I've seen something like that in our grate". It was a bit like cinders from a coalfire.[54]

Similarly, a member of staff at the time remembers the Director describing the sample as 'a little bit of cigarette ash'.[55] Even Neil Armstrong had described moon soil as 'powdered charcoal'.[56] Many visitors were nonplussed by its appearance: one later admitted, 'it meant nothing to me … I was more into the tatty old mummy they had on display'.[57]

Nevertheless, the exhibition attracted a staggering number of visitors, for whom the rock, however unappealing, had special meaning. Queues formed along Oxford Road, and by the end of the five-day exhibition 24,000 people had filed past the rock.[58] Through travel and ritual, one of 300 samples was rendered singular and iconic. Even though there have since been many such samples on display from the Apollo missions – perhaps most famously from Apollo 17 at the Smithsonian Institution – each one is imbued with distinct singularity when on display as *the* moon rock.[59]

After the exhibition, Zussman's team returned to work on the samples, the object returned from display in a museum space to analysis in the laboratory space.[60] Zussman and MacKenzie attended the

Lunar Science Conference in Houston in early January 1970, at which NASA distributed samples from Apollo 12, and the investigators went back to work again.[61] Zussman continued to lecture at the Museum, and sample 10085,2 remained in the civic memory of Manchester. Meanwhile, every surviving speck of the sample itself was accounted for and returned to Houston, to join the collection of lunar material that NASA has since made available for research and teaching loans.

Conclusion

Museum objects enable scientific collaboration. In doing so, they are not stable entities over time and space, but their meanings (and physical integrity) shift as they change hands, and as different actors impact upon their trajectories. These three objects have apparently little in common. But in tracing their biographies, some conclusions can be drawn about the status of material culture and the role of museums in twentieth-century science, and about research culture in Manchester.

None of the items was collected by a curator. Rather, they were found by diverse individuals with varied reasons for being where they were, and different intentions when gathering the objects. Frederick Townsend, Erfai (or the unnamed boy) and Neil Armstrong were all first links in contingent collecting chains. At the time they found the objects they were all in a way being paid to pick things up – cables, relics and moon rocks. Although Townsend gathered shells outside his professional remit to further his own interests, he nonetheless peeled them from the cable as part of a system of exchange, thus fulfilling an agreement with Melvill, as Erfai was to Mackay (and thence to Petrie), and Armstrong to NASA. The immediate collectors gathered molluscs, mummies or moon rocks in order to satisfy a patron or employer (thereby promoting their own interests and careers).

All these transactions were undertaken in the course of imperial projects, geographic or lunar. By collecting natural and cultural objects from colonies and protectorates, the British strengthened their imperial infrastructure in the decades around 1900.[62] Similarly, the space race was part of the competition between twentieth-century superpowers. Iconic objects such as mummies and moon rocks contributed to prestige projects undertaken by quasi–imperial organisations. The personnel involved were at the periphery specifically to collect (as was Petrie), had collecting as part of their remit (Armstrong) or were collecting 'on the side' (Townsend). They were technical explorers, maintaining the imperial infrastructure on vessels such as the *Patrick Stewart* and Apollo 11 with their specific training and skills (dredging, surveying or moon walking).

The status of each patron reflects not only the importance of empire but the changing character of scholarship and collecting over the twentieth century. Townsend's specimen went to James Cosmo Melvill, a wealthy individual with personal interest in natural history. As a powerful amateur his wealth and autonomy placed him at the centre of a web of acquisition that spanned continents. As Chair of the Museum committee, Melvill was engaged in a reciprocal relationship with the institution. He bestowed specimens on the Museum, and the professional Standen – who himself had first practiced conchology as an amateur – worked as his subordinate collaborator. But the relative standing of amateur collector, Museum committee and curator was shifting. Although Jesse Haworth (and his money) played a key role in bringing Khnum-Nakht to Manchester, it was Flinders Petrie, an autodidact but nonetheless professional scholar and collector, who is most prominent in the biography of item 11725. The tomb of the two brothers was important to Petrie not only as an archaeological find, but also as a financial resource to further his collecting, and thereby his career. Nevertheless, this is not a story of unfettered professional ascendance. Amateur expertise and benevolence continued to be vital to museum activity throughout the twentieth century.[63]

The collecting route of the moon rock stands in marked contrast. Nowhere in the object biography of lunar sample 10085,2 did an amateur scientist play a significant part.[64] Even the very act of gathering material culture was but a tiny part of the vast scientific enterprise that allowed humans to walk on the moon. Armstrong was not collecting on behalf of an individual, but for a massive technoscientific institution with enormous government support. The specimen arrived at Manchester not through informal collecting networks, but via national research agencies.

This much was very different. In common with *Conus clytospira* and Khnum-Nakht, however, the moon rock was studied by a group of interested individuals upon arrival in Manchester. The specimens were returned from the far reaches of empire for uses that reflected the status of the Museum at the time. Classification, analysis and display were all to varying extents undertaken in the Museum and the University. In considering these practices, it is revealing to consider not only who undertook work on the objects, but *where*. Shells were sifted, identified and arranged by the Manchester conchologists, led by Melvill and including a number of curators as his junior partners, and much of this work was undertaken in his private residence – science enacted in the semi-domestic space.[65] These same curators worked side by side with rising professional Margaret Murray as she unwrapped Khnum-Nakht, drawing on expertise from across the University and from London. Although the unwrapping was not carried out in the Museum, the Chemistry Lecture Theatre was effectively its satellite in this respect. The museum has endured as a site for research far beyond the rise of the laboratory in late-nineteenth century science.[66] At the other end of the century, it was a professional academic curator who led the mummy team, and although the two brothers' remains and those of mummy 1770 were analysed in laboratory spaces across campus, the activities still centred on the museum. Acting as a conduit for a bundle of relationships, the Manchester Museum channelled informal associations established by word of mouth, chance introductions and lunchtime conversations in the Senior Common Room, where keepers rubbed shoulders with other University staff. Museum personnel taught in University departments, and some, such as Standen and Garner, had worked there. These connections extended further, across the globe and beyond, but they intersected in Manchester through the work carried out on the objects.

Even the moon rock's appearance in the Museum was dependent on informal relations and serendipity. Only with Zussman's move from Oxford to Manchester, his personal contact with Eagar and interest in the Museum did the sample come to be exhibited in the city. But the research surrounding the moon rock did not involve Eagar or Rushton or even the Museum. Rather, 'big science' was carried out under lock and key in the laboratory. Not that the Museum was insignificant: the television cameras and tens of thousands of visitors (however unimpressed) speak volumes about the demonstrative and performative character of science, even in the twentieth century. This was also evidenced by the mummy unwrappings at both ends of the century, and the efforts undertaken by Murray, David and Zussman to employ the mass media and engage with the public. At all stages, the serious, scientific character of the work was emphasized: the precise location of *clytospira*; the autopsy instruments used to unwrap 1770; and the projection of the moon sample intended to allow the visitor to view it as a scientist would. This was science as civic performance, in the early and late twentieth century alike, and the museum was a stage.

The mollusc and the moon rock indicate a broader twentieth-century shift from museum as a site for scientific research to a site for its promulgation. The mummy work, however, demonstrates that a museum could still be a site for both, especially in areas involving interdisciplinary collaboration. Any shift in status from museum to laboratory, from amateur to professional, did not follow strict linear chronology. Petrie the professional and Melvill the amateur were operating contemporaneously, and the moon rocks were analysed in a laboratory while mummies were studied in a museum. Different sites, practices and ways of knowing overlapped and interconnected.[67]

Whatever this study may reveal about the status of different individuals and institutions, at its core this has been a study of objects. As they moved through (outer) space, the objects took on different meanings. Like other forms of object-people interaction, scientific research conferred new status and categories on the objects. They became different things, from an ornamental shell to a type specimen, from grave booty to pathological evidence, from the dust that might have been stuck to an astronaut's shoe to the centre of a thriving exhibition. Here, the 'mummy' stands out as distinct by virtue of its human origin. But despite the powerful status of the corpse, Khnum-Nakht nonetheless followed a similar trajectory and classification processes to other objects in the collection. As the remains of a man of a different time and ethnicity his cadaver was somehow less human than its counterparts on dissecting tables elsewhere in the Medical School.

The (living) individuals involved benefited from working with the objects, gaining in prestige, advancing their career or accruing monies. So too with the objects, which through their interactions with people accrued value and fame, and derived identity from the people with whom they become associated. The mollusc became *Conus clytospira* Melvill & Standen (however briefly before becoming a junior synonym); Khnum-Nakht became forever associated with Flinders Petrie, Margaret Murray and later the mummy team; and 10085,2 was so popular partly because it had been gathered by Neil Armstrong. They had different meanings for different practitioners and communities. They all, however, demonstrate a vibrant and dynamic research culture in twentieth-century Manchester.

Notes

1. S. J. M. M. Alberti, 'Placing nature: natural history collections and their owners in nineteenth-century provincial England', *British Journal for the History of Science*, 35 (2002), pp. 291–311. Thanks to Fay Bound Alberti, Ben Alberti, Rosalie David, Roy Garner, David Green, Nicola Gower-Jones, Henry McGhie, John Pickstone, John Prag, Christina Riggs and Jack Zussman. The similarity in title to W. T. Alderson (ed.), *Mermaids, mummies, and mastodons: the emergence of the American museum* (Washington, D.C.; 1992) is not intentional.

2. There is a well established literature on object biographies across a number of disciplines. See for example S. J. M. M. Alberti, 'Objects and the museum', *Isis*, 96 (2005), pp. 559–71; A. Appadurai (ed.), *The social life of things: commodities in cultural perspective* (Cambridge, 1986); C. Gosden and Y. Marshall, 'The cultural biography of objects', *World Archaeology*, 31 (1999), pp. 169–78.

3. J. C. Melvill and R. Standen, 'Description of *Conus (Cylinder) clytospira* sp. n. from the Arabian Sea', *Annals and Magazine of Natural History*, 4

(1899), pp. 461–3; J. C. Melvill, 'Note on *Conus clytospira* Melv. and Stand.', *Proceedings of the Manchester Literary and Philosophical Society*, 44 (1899), pp. x–xi; F. W. Townsend, 'Notes on shell collecting in the northern parts of the Arabian Sea, including the Gulfs of Oman and Persian Gulf in the years 1890–1914', *Proceedings of the Malacological Society*, 18 (1928), pp. 118–26.

4. F. W. Townsend's annotated copy of J. C. Melvill and R. Standen, 'The mollusca of the Persian Gulf, Gulf of Oman and Arabian Seas as evidenced through the collections of Mr. F. W. Townsend, 1893–1900, with descriptions of new species. Part 1. Cephalopoda, Gastropoda and Scaphopoda', *Proceedings of the Zoological Society of London* (1901), pp. 327–460; and J. C. Melvill, 'Descriptions of thirty-four species of marine mollusca from the Arabian Sea, Persian Gulf and Gulf of Oman', *Manchester Memoirs*, 41 (1897), pp. 5–26, both in the Manchester Museum Zoology Archive.

5. A. Trew, *James Cosmo Melvill's new molluscan names* (Cardiff, 1987); J. C. Melvill, *A brief account of the General Herbarium formed by James Cosmo Melvill, 1867–1904; and presented by him to the Museum in 1904*, Manchester Museum Publications, 54 (Manchester, 1904).

6. S. P. Dance, *A history of shell collecting* (London, 2nd ed., 1986); J. M. Chalmers-Hunt (ed.), *Natural history auctions 1700–1972: a register of sales in the British Isles* (London, 1976).

7. A. Norris, 'The Conchological Society of Great Britain and Ireland: the early years', *The Naturalist* (1982), pp. 131–4.

8. J. W. Jackson, 'Obituary notice: Robert Standen', *Journal of Conchology*, 17 (1925), pp. 41–5; *Report of the Manchester Museum* (1890–4), Manchester Museum Central Archive [hereafter, 'MMCA'].

9. R. Standen, 'Lancashire land and freshwater mollusca', *The Naturalist* (1887), pp. 155–76.

10. On taxonomic and morphological work in Manchester biology at the museum and elsewhere in Owens College, see A. Kraft and S. J. M. M. Alberti, '"Equal though different": laboratories, museums and the institutional development of biology in the late-Victorian industrial north', *Studies in History and Philosophy of Biological and Biomedical Sciences*, 34 (2003), pp. 203–36.

11. J. C. Melvill, 'Upon the principles of nomenclature, and their application to the genera of recent mollusca', *Journal of Conchology*, 8 (1897), pp. 435–79.

12. J. C. Melvill, 'Notes on the subgenus *Cylinder* (Montfort) of *Conus*', *Memoirs of the Manchester Literary and Philosophical Society*, 10 (1887), pp. 76–90; Melvill, 'Note on *Conus*', pp. x–xi.

13. J. C. Melvill and R. Standen, *The marine mollusca of Madras and the immediate neighbourhood. Notes on a collection of marine shells from Lively Island, Falklands; and other papers*, Manchester Museum Publications, 24 (Manchester, 1898); J. C. Melvill and R. Standen, 'Report on the marine

mollusca obtained during the first expedition of Prof. A. C. Haddon to the Torres Straits in 1888–89', *Journal of the Linnean Society of London – Zoology*, 27 (1899), pp. 150–206; J. C. Melvill and R. Standen, *Catalogue of the Hadfield collection of shells from the Lifu and Uvea, Loyalty Islands*, Manchester Museum Owens College Handbooks (Manchester, 2 vols, 1895–7); J. C. Melvill and R. Standen, 'The marine mollusca of the Scottish National Antarctic Expedition', *Transactions of the Royal Society of Edinburgh*, 46 (1907), pp. 119–57; J. C. Melvill and R. Standen, 'The marine mollusca of the Scottish National Antarctic Expedition' [second instalment], *Transactions of the Royal Society of Edinburgh*, 48 (1912), pp. 333–66.

14. M. A. Murray (ed.), *The tomb of two brothers*, Manchester Museum Museum Handbooks, 68 (Manchester, 1910); W. M. F. Petrie, *Gizeh and Rifeh*, British School of Archaeology in Egypt and Egyptian Research Account, 13 (London, 1907).
15. W. R. Dawson and E. P. Uphill (eds.), *Who was who in Egyptology* (London, 1951), rev. M. L. Bierbrier (London, 3rd ed., 1995).
16. M. S. Drower, *Flinders Petrie: a life in archaeology* (London, 1985).
17. Manchester Museum Committee Minutes, 4 July 1907, MMCA; W. E. Hoyle, 'Preface', in Murray, *The tomb of two brothers*, p. 3.
18. Manchester Museum Committee Minutes, 28 Oct. 1907, MMCA.
19. Manchester Museum Department of Archaeology Accessions Register, 4 vols, Manchester Museum Egyptology Archives.
20. S. Pearce, R. Flanders, M. Hall and F. Morton (eds.), *The collector's voice: critical readings in the practice of collecting*, vol. 3: *Imperial voices* (London, 2002), pp. 299–306.
21. *Manchester Courier*, 8 May 1908, Manchester Museum news cuttings ledger 13 Feb. 1905 – 11 Oct. 1935, p. 47, MMCA.
22. *Manchester Evening News*, 7 May 1908, Manchester Museum news cuttings ledger 13 Feb. 1905 – 11 Oct. 1935, p. 48, MMCA.
23. *Daily Dispatch*, 9 May 1908, Manchester Museum news cuttings ledger 13 Feb. 1905 – 11 Oct. 1935, p. 48, MMCA.
24. *Manchester Courier*, 8 May 1908, Manchester Museum news cuttings ledger 13 Feb. 1905 – 11 Oct. 1935, p. 47, MMCA.
25. Murray, *The tomb of two brothers*.
26. J. Cameron, 'The anatomy of the mummies', in Murray, *The tomb of two brothers*, pp. 33–47.
27. Besides the work undertaken on the two brothers, the Australian anthropologist Grafton Elliot Smith undertook Egyptological studies during his tenure at the University from 1909 as Professor of Anatomy and later Dean of Medicine. Smith formulated a diffusionary hypothesis for the spread of mankind that originated in Egypt. See J. S. B. Stopford, 'The Manchester period', in W. R. Dawson (ed.), *Sir Grafton Elliot Smith: a biographical record by his colleagues* (London, 1938), pp. 151–65; H. A. Waldron, 'The study of the human remains from Nubia: the contribution of

Grafton Elliot Smith and his colleagues to palaeopathology', *Medical History*, 44 (2000), pp. 363–88.
28. *Manchester Museum Annual Report* (1972–3), MMCA; R. Garner, interview by the author, MD recording, 24 Nov. 2004, *Re-collecting at the Manchester Museum* Oral History Project [hereafter '*RMM*'], disc S3 track 7, MMCA.
29. R. Garner, interview by the author, MD recording, 24 Nov. 2004, *RMM* disc S3 track 7, MMCA.
30. A. Curry, 'The insects associated with the Manchester mummies', in A. R. David (ed.), *The Manchester Museum mummy project* (Manchester, 1979), pp. 113–17. For further Museum staff involvement, see C. W. A. Pettitt and G. Fildes, 'The international Egyptian Mummy Data Base', in A. R. David (ed.), *Science in Egyptology* (Manchester, 1986), pp. 174–81.
31. David, *Science in Egyptology*; P. Lambert-Zazulak, 'The International Ancient Egyptian Mummy Tissue Bank at the Manchester Museum', *Antiquity*, 74 (2000), pp. 44–8.
32. Manchester Museum Committee Minutes, 20 May 1974; 3 June 1974, MMCA.
33. E. Tapp, 'The unwrapping of a mummy,' in David, *The Manchester Museum mummy project*, pp. 83–93.
34. A. R. David, interview by the author, MD recording, 11 Apr. 2005, *RMM* disc S13 track 4, MMCA.
35. 'Revelations of a young mummy', *Guardian*, 11 June 1975, p. 1.
36. 'Under wraps: Rosalie David in conversation', *Archaeology Online*, 6 Feb. 2001.
37. A. R. David, interview by the author, MD recording, 11 Apr. 2005, *RMM* disc S13 track 6, MMCA.
38. F. Filce Leek, 'The dental history of the Manchester mummies,' in David, *The Manchester Museum mummy project*, pp. 65–77; Cameron, 'The anatomy of the mummies'.
39. E. Tapp, A. Curry, and C. Anfield. 'Sand pneumoconiosis in an Egyptian mummy', *British Medical Journal*, 3 May 1975, p. 276; David, *The Manchester Museum mummy project*, pp. 25–64, 95–111.
40. R. A. H. Neave, 'The reconstruction of the heads and faces of three ancient Egyptian mummies', in David, *The Manchester Museum mummy project*, pp. 149–57. See also R. Neave and A. J. N. W. Prag, *Making faces: using forensic and archaeological evidence* (London, 1997).
41. David, *The Manchester Museum mummy project*, p. vii.
42. R. Garner, interview by the author, MD recording, 24 Nov. 2004, *RMM* disc S3 track 7, MMCA.
43. P. Ulivi and D. M. Harland, *Lunar exploration: human pioneers and robotic surveyors* (London, 2004); R. Turnill, *The moonlandings: an eye-witness account* (Cambridge, 2003); W. A. McDougall, *The heavens and the Earth: a political history of the Space Age* (New York, 1985); 'Apollo 11', http://

nssdc.gsfc.nasa.gov/planetary/lunar/apollo11info.html, accessed 22 Feb. 2005.

44. J. Arnold et al, 'Summary of Apollo 11 Lunar Science Conference', *Science*, 167 (1970), pp. 449–51.

45. D. H. Anderson to J. Zussman, telegram, Sep. 1969; 'The first lunar samples arrive in Britain', Science Research Council press release, 19 Sep. 1969, both in Jack Zussman personal collection [hereafter 'JZPC'].

46. J. Zussman, interview by the author, MD recording, 4 Mar. 2005, Manchester, *RMM* disc S8 track 2, MMCA.

47. J. Zussman, 'Security plans for lunar samples', MS, Sep. 1969, JZPC.

48. J. Zussman, 'Proposed investigation of lunar surface material', application, Oct. 1967, JZPC; J. C. Bailey, P. E. Champness, A. C. Dunham, J. Esson, W. S. Fyfe, W. S. MacKenzie, E. F. Stumpfl and J. Zussman, 'Mineralogical and petrological investigations of lunar samples', *Science*, 167:3918 (1970), pp. 592–4.

49. J. Zussman, interview by the author, MD recording, 4 Mar. 2005, Manchester, *RMM* disc S8 track 3, MMCA.

50. 'Dr R. Michael C. Eagar (1919–2003), formerly Keeper of Geology', *Manchester Museum Annual Report* (2003), p. 28, MMCA.

51. J. Zussman to R. M. Payne, 24 Sep. 1969; R. M. Payne to J. Zussman, 25 Sep. 1969; R. M. Payne to J. Zussman and D. E. Owen, 26 Sep. 1969, JZPC.

52. *Manchester Museum Annual Report* (1969–70), pp. 6–7, MMCA.

53. R. M. C. Eagar and D. E. Owen, 'Moon rock in Manchester', *Museums Journal*, 69 (1970), pp. 159–60.

54. J. Zussman, interview by the author, MD recording, 4 Mar. 2005, Manchester, *RMM* disc S8 track 3, MMCA.

55. A. J. N. W. Prag, interview by the author, MD recording, 17 Nov. 2004, *RMM* disc S1 track 11, MMCA.

56. Quoted in Turnill, *The moonlandings*, p. 262.

57. N. Gower-Jones, interview by the author, MD recording, 17 Mar. 2005, *RMM* disc V1 track 2, MMCA.

58. R. M. C. Eagar, 'The moon in a geological gallery', *Museums Journal*, 64 (1964), pp. 132–7.

59. See a discussion of the Smithsonian moon rock in D. J. Meltzer. 'Ideology and material culture', in R. A. Gould and M. B. Schiffer (eds.), *Modern material culture: the archaeology of us* (London, 1981), pp. 113–25.

60. They were mostly basalts, which are fine-grained igneous rocks (that is, solidified from molten lava); breccias, which are compacted or welded together aggregates of rock and mineral fragments; and the so-called 'soil' of very tiny crystals and glassy spheres. The team employed electron and light microscopy and X-ray diffraction to identify and compare the composition of the lunar materials, and from there to suggest mechanisms of formation for 10044. J. C. Bailey, P. E. Champness, A. C. Dunham,

J. Esson, W. S. Fyfe, W. S. MacKenzie, E. F. Stumpfl and J. Zussman. 'Mineralogy and petrology of Apollo 11 lunar samples', in A. A. Levinson (ed.), *Proceedings of the Apollo 11 Lunar Science Conference, Houston, Texas, 5–8 January 1970* (New York, 3 vols, 1970), vol. 1, pp. 169–94.

61. For example, F. G. F. Gibb, E. F. Stumpfl and J. Zussman, 'Opaque minerals in an Apollo 12 rock', *Earth and Planetary Science Letters*, 9:3 (1970), pp. 217–24; F. G. F. Gibb and J. Zussman, 'Zoned olivines in 4 Apollo 12 samples', *Earth and Planetary Science Letters*, 11:2 (1971), pp. 161–7.

62. T. Barringer and T. Flynn (eds.), *Colonialism and the object: empire, material culture and the museum* (London, 1998); F. R. Myers (ed.), *The empire of things: regimes of value and material culture* (Oxford, 2001).

63. On the status of amateurs in late-nineteenth century provincial natural history, see S. J. M. M. Alberti, 'Amateurs and professionals in one county: biology and natural history in late-Victorian Yorkshire', *Journal of the History of Biology*, 34 (2001), pp. 115–47.

64. This is not to ignore the contribution of amateur astronomers in general over the twentieth century.

65. S. D. Chadarevian, 'Laboratory science versus country house experiments: the controversy between Julius Sachs and Charles Darwin', *British Journal for the History of Science*, 29 (1996), pp. 17–41.

66. Kraft and Alberti, '"Equal though different"', pp. 203–36.

67. J. V. Pickstone, *Ways of knowing: a new history of science, technology and medicine* (Manchester, 2000).

ARCHIVES AND LIBRARIES
The Collections Centre at the Museum of Science and Industry in Manchester

Jan Hargreaves

In 1990, Museum archivist Ann Jones and her colleague Elizabeth Sprenger wrote an article about the Library and Record Centre, as it was then known, for volume IV of the *Manchester Region History Review*. It is 17 years since that article was written, and many changes have taken place. The Museum has opened a number of new galleries, including the Textile Gallery and the Manchester Science Gallery in the Main Building. The 1830 Warehouse has been transformed from an almost derelict shell to a living exhibit housing the Gas Gallery, Underground Manchester and will soon include the Communications Gallery. Entry to the Museum is now free and visitor figures have risen from the low 100,000s to around half a million.

The biggest change for the archive collections is that the Library and Record Centre is now the Study Area, and forms one part of the Collections Centre. The Collections Centre opened in September 2001 and occupies the basement of the main Museum building on Lower Byrom Street. The basement was completely refurbished thanks to funding through the Heritage Lottery Fund, as part of the final phase of developing the Museum site. The purpose of the Collections Centre is to bring together some of the Museum's reserve collection of objects and make them available to visitors through the use of mobile storage and open-plan store rooms. The Collections Centre also enables the provision of better access to the archive collections.

Large objects are on display in glass-fronted stores housed in the Colonnades, alongside the lower car park. There is further storage for large objects inside the basement area, along with a run of mobile storage units. The Museum's smaller objects are displayed in these mobile units. Visitors can view objects that used to be shut away in remote access storage by opening the drawers of these units. Special trails help the curious to learn more about these objects, and curatorial staff provide guided tours of the large object stores. The Collections Centre also includes a handling area which is currently used for staff-led handling sessions. During these sessions Museum staff explain what some of the objects in the Museum's collections might have been used for and why we keep them.

In the Welcome area there is a small library of journals, periodicals and general-interest reference books. There are also books suitable for use with young children, and a collection of information files for the casual browser. Copies of subject information sheets and company histories that relate to the object and archive collections held at the Museum are available, and a member of staff is on hand to answer questions.

To help make more people aware of the archive collections, the Collections Centre staff put on quarterly exhibitions of items from the collections in the archive display case. Exhibitions so far, since the display case was installed, have been in celebration of the Beyer, Peacock locomotive company; the more unusual archives we hold among our collections; and items that record pioneering achievements. Visitors are welcome to open the drawers and read the literature provided, and are encouraged to come into the Study Area to carry out further research.

As well as the physical changes to the former Library and Record Centre, a considerable number of new archive collections have been donated to the Museum since Ann Jones and Elizabeth Sprenger wrote their article. In 1990, the Museum held around 500 archive collections, brought in since 1965. In the past 17 years, the number of archive collections has increased to around 1,250. This represents more than double the number of collections in just over half the time. By 2004, the archive strongroom was almost full, with roughly 1300 linear metres of documents. We have installed additional mobile shelving to allow for future expansion of our archive collections, and in the last three years approximately 200 linear metres of new archive material have been added to our holdings.

Over the past 10 years, the Museum has accepted three of its larger collections. The Ferranti company archive [1] occupies 230 linear metres of shelving, and documents the running of the company through board minutes, research and development records, and archives drawn from departments and sub-divisions of the company. The largest department was Strategic Services, the records of which include advertising material and product brochures, as well as documents relating to recruitment, training and pensions. The Linotype and Machinery archive [2] runs to 60 linear metres of documents, and tells the story of the first company to supply hot metal-typesetting equipment to the printing industry. Publisher Joseph Lawrence formed the company in 1889 and supplied linotype machines around the world. The company had branches in Africa, the Far East and on the Indian sub-continent. The archive includes sales registers, accounts, trade literature and manuals for the different types of machine that the company produced. The third major collection, acquired in 2005, is the Benger's Food archive,[3] which consists of 36 linear metres of advertising copy, product packaging, photographs, correspondence with clients around the world, and details of the scientific testing of products. Benger's was established in Manchester at the time of the Crimean War, manufacturing a food supplement for returning soldiers. The product was developed during the nineteenth and early-twentieth centuries to include dietary supplements for babies, children and the elderly. The company moved to Holmes Chapel in the 1930s and was later taken over by Fison's. Production switched away from food supplements to a range of pharmaceutical products.

Advertisement for Benger's Food from an album of advertising proofs c.1910. Collection reference 2005.84, reproduced with the permission of the Senior Archivist, Museum of Science and Industry in Manchester

Not all of the archive collections we accept are so large. Most are small collections, often single items, acquired in support of gallery development, or as a result of a large collection being donated. For example, since the acquisition of the main Ferranti Company archive in 1996, the Museum has received supplementary collections of Ferranti material on a regular basis. In 2002 the Museum acquired the personal papers of the Ferranti family.[4] The collection consists of the private letters of Sebastian Ziani de Ferranti, including those written to his mother while he was at boarding school, as well as business correspondence and diaries, lecture notes, and notes relating to the development of new products and processes. The letter books and diaries are interspersed with diagrams and sketches of ideas S. Z. de Ferranti was working on at the time. It is an exciting collection which documents the early stages of the company's development.

Another archive, acquired in 2005, which links to the first archive collection accepted by the Museum in 1966, is that of Herbert W. Garratt. Garratt was a locomotive engineer who worked with Beyer, Peacock to produce the Beyer-Garratt engine. The archive, which was obtained through the auspices of a former Director, Dr R. L. Hills, includes Garratt's apprenticeship indenture, correspondence written during his travels abroad as an engineer, and documentation relating to negotiations with Beyer, Peacock regarding royalty payments for the Beyer-Garratt.[5]

The Museum currently averages 30 new archive collections per year, although some years are busier than others. In 1996 the archive accepted some 350 new collections. These included a number of archives relating to the work of the Metropolitan Vickers Electrical Company (Metrovicks) and its successors AEI and GEC Alsthom.[6] One of these collections[7] was rescued by Ian Artus, an employee at GEC Alsthom, and consists of detailed correspondence relating to the development of internal combustion turbines by Metrovicks between 1937 and 1944. The R. R. Whyte Collection[8] similarly documents the production of gas turbines at Trafford Park from the 1940s to the 1970s.

A collection with strong links to the R. R. Whyte collection is the Frank Harris Collection.[9] Harris was an employee at Metrovicks and later AEI, working in the field of gas turbines. The collection documents his career in gas turbine technology development, and makes reference to R. R. Whyte.

Another of the collections acquired during 1996 is the Owen Ashmore Collection of photographs, publications, maps and slides relating to industrial archaeology in Lancashire and Cheshire.[10] Ashmore was a founder member of the Manchester Region Industrial Archaeology Society, and led the Industrial Archaeology group of the Stockport Historical Society as well as teaching on and directing the University of Manchester Extra Mural Department's programme of industrial archaeology courses.

Other collections brought in that year were a notebook containing hand-written notes and printed articles and agreements on working conditions in the printing industry,[11] lantern slides, negatives and photographic prints of machinery manufactured by Joseph Adamson & Co.,[12] and architectural drawings of the Lower Byrom Street Warehouse, now the Main Building of the Museum, received from British Rail.[13]

Complementing the photographic collection referred to above, Joseph Adamson & Co. donated its business archive to the Museum in 2003.[14] The collection consists of journals, drawings, correspondence, reports, plans, manuals, advertisements, brochures, printing blocks, newspaper cuttings and photographs dating from the 1870s to the 1990s. Joseph Adamson, nephew of Daniel Adamson, founded the company with Henry Booth in 1874. Adamson had worked for his uncle's business, Newton Heath Boiler Works, from the time of his apprenticeship in 1858 to his role as commercial and technical manager between 1867 and 1873. From the 1890s, the company produced cranes as well as boilers, although their main product remained the 'Lancashire' boiler.

As mentioned above, we collect archive material in order to support gallery developments. In 1997, the Museum opened its Textile Gallery in the Main Building. Prior to its opening, the Museum sought archive collections that would support the gallery's development. Key collections included the Paterson Zochonis archive of fabric samples for the West African market[15] and the Dimoldenberg Collection of textiles, design samples and drawings.[16] The Museum acquired other textile-related collections after the opening of the gallery, including the Coates-Viyella Collection of records relating to the research, development and patenting of Terylene.[17]

In order to support the opening of the Collections Centre, the Museum acquired the Pollock Collection of clay pipes in April 2001. The objects came with a substantial archive that includes advertising material, catalogues, photographs, price lists and business correspondence.[18] The Pollock family manufactured and distributed clay pipes throughout the twentieth century, selling them as far afield as Australia.

Since the Museum took over the running of the Air and Space Hall in 1985, our collecting has included aviation archives. Shortly after the 1990 article was published, the Museum accepted a collection of papers relating to Roy Chadwick, aircraft designer with A. V. Roe.[19] The collection includes documents relating to awards received by Chadwick, photographs, and papers relating to the '244' aircraft built by A. V. Roe. A similar acquisition is the A. C. Jack Collection of photographs, handbooks and certificates relating to his career with A. V. Roe.[20]

Paterson Zochonis shipping label showing the Slipper Mark. Collection reference MS Papers 0424, reproduced with the permission of the Senior Archivist, Museum of Science and Industry in Manchester

The Museum has sought to broaden its collecting in recent years, and this is reflected in the types of archive collection we have acquired. In 1999 we received the Factory Communications Ltd archive from former Factory director Rob Gretton.[21] The collection documents the development of the Haçienda nightclub through design plans by

Ben Kelly, guest lists, financial records for club nights, and papers relating to its subsequent closure. The collection also includes artwork for various recordings released on the Factory record label, designed by Peter Saville Associates, and posters advertising Haçienda club nights, along with documentation revealing the financial management of the company.

In 2002, the Museum acquired the object and archive collections of the former Lancashire Coal Mining Museum, based in Salford.[22] The Mining Museum was forced to close due to lack of funds, and we agreed to take the object collection and archive. The archive had to be split since the majority of it consisted of public records. The Museum is not a repository designated by The National Archives for the preservation of public records, and so we transferred all the items relating to coal mines and coal mining companies to the Lancashire Record Office in Preston. The Museum has retained items of a private nature, such as union records, memorabilia and employee reminiscences, along with a sizeable research library. Among the archive papers are records relating to the archaeological study of the Wet Earth Colliery at Croal Irwell. The colliery was established between 1751 and 1756. After it ceased production as a mine, the tunnel system was used for the supply of water until 1960. The documentation of the archaeological survey includes photographs and reports, minutes and correspondence.

During 2005, new donors brought a wide range of archive collections into the Museum. Early in the year, we received Ken Johnson's research archive of notebooks and correspondence, covering the period when he worked at Ferranti, ICL and Manchester University.[23] Mr Johnson worked primarily as an electronic circuit designer, and the collection includes samples of the circuitry he designed for Ferranti computers. At the other end of the scientific scale is the J. Jordan Celluloid Heel collection.[24] J. Jordan was an inventor who developed a method of illuminating shoe heels in the 1940s. The product was intended for the dance market, to be worn by professional dancers during performances. The collection has photographs of the illuminated heels and details of the patent awarded to Mr Jordan.

All of the collections mentioned in this article, and more, are available for researchers to make use of. You can obtain more information on our archive holdings by contacting the Senior Archivist on 0161 606 0127 or by emailing archive@msim.org.uk. The Study Area is open to the public every Tuesday to Thursday from 10 a.m. to 4.30 p.m. Appointments are not always necessary. The Collections Centre is open from Tuesday to Saturday at the same times.

Notes

1. Collection reference 1996.10.
2. 1997.20.
3. 2005.84.
4. 2002.22.
5. 2005.60.
6. 1996.139, 1996.530, 1996.1735, 1996.2339 and 1996.3235.
7. 1996.3235.
8. 1996.530.
9. 1999.39.
10. 1996.238.
11. 1996.2132.
12. 1996.531.
13. 1996.3037.
14. 2003.54.
15. 1995.2.
16. 1996.35.
17. 1998.15.
18. 2001.281.
19. 1991.630.
20. 1999.88.
21. 1999.16.
22. 2002.19.
23. 2005.57.
24. 2005.71.

MUSEUMS

Making Manchester massive: a missed opportunity?

Exhibition review: 'Manchester Science: discoveries that changed the world', Manchester Museum of Science and Industry

Francis Neary, CHSTM, University of Manchester

The Manchester Museum of Science and Industry is well known for its galleries covering the Manchester textile industry and steam power in the Victorian city, its transport galleries filled with large objects, and with its interactive exhibits for children. However, scientific research and discovery in Manchester was under-represented until a new permanent gallery 'Manchester Science: discoveries that changed the world' opened in 2004.

On entering the gallery you are confronted with the responses of past visitors. A long wall is devoted to index-style cards on which visitors have answered a series of questions about science in Manchester and scientific development more generally. Some questions are on controversial topics like animal experimentation and global warming, some are intended to fuel the imagination, for instance: 'How could science make Manchester a better place?' and 'If you could make a scientific discovery, what would it be?' The responses are a measure of the exhibition's success in inspiring visitors and communicating Manchester's place in scientific research: they range from the ingenious paragraph to the banal one-liner; they suggest multiple points of view and encourage critical thought about the exhibition. But it is a pity that this beginning is also where the interaction ends, the rest of the exhibition being much more passive and lacking opportunities for visitor participation.

The centre of the exhibition is dominated by a long display with a timeline on each side. One side looks at science in the media, and the other places Manchester innovations in the context of a world science timeline, but cleverly, some of the objects can be viewed from both sides so they appear in two different contexts. The 'Science

Entering the gallery

Visitor feedback wall (left) and 'Science in the Media' timeline (right)

in the Media' timeline is really a series of blown-up newspaper cuttings interspersed by objects. It is very dense in text and when I was there few of the visitors stopped to read the details. The only video installation in this area depicts the Apollo 11 moon landing, but it is not here connected to Manchester (see the article in this volume by Sam Alberti). In fact it is accompanied by a *Daily Express* article by Sir Bernard Lovell on the earlier unmanned moon landing by the Soviet Lunar 9 rocket in 1966. More generally, some of the media articles are from the Manchester press or about Manchester science, while

others show how Mancunians might have read about science done elsewhere. But was there not enough scientific research going on in Manchester from 1750 to the present to devote this section to stories about Manchester science without being parochial? Perhaps an opportunity has been missed, at this early stage in the gallery, to convey the global importance and connections of Manchester science.

The cases of smaller objects in the timeline were all made in Manchester and they are complemented by three large objects: a 1974 mass spectrometer made by VG Micromass, Wythenshawe; a 1960s electron microscope made by Associated Electrical Industries based in Trafford Park, and an inorganic chemistry bench from the Manchester Technical College (the forerunner of UMIST). The large objects make an impressive display but the labelling hardly enhances the Manchester context. We learn that the mass spectrometer was bought by the University of Oxford to study the effects of marijuana on the human body, and that the electron microscope went to Birmingham Medical School and later to Withington Hospital in Manchester. However, we discover little about the Manchester companies that made these fascinating objects. This failure to fully develop the Manchester context is also apparent in the labelling of the small objects in the timeline. An excellent range of well displayed objects can be found here, as diverse as differential analysers, surgical instruments, safety lamps, light bulbs and memorabilia from the opening of the Liverpool and Manchester Railway in 1830. But again the interpretation fails to tell the Manchester stories behind the objects.

In moving from the media timeline to the world science timeline, a section called 'Boomtown' consists of one panel and an art installation. The panel tells us how, around 1800, Manchester was a thriving city whose population had tripled in 30 years to provide a workforce for over 100 cotton spinning firms. That this section is so tiny suggests a weak appreciation of the links between

1960s electron microscope made by AEI in Trafford Park

Early 1900s inorganic chemistry bench from Manchester Technical College

The Davy safety lamp made in Eccles

Manchester's scientific heritage and its industry. For instance, Joseph Whitworth's precise measuring instruments are displayed but with no reference to their importance in standardising the manufacture of the tiny fittings required to make accurate scientific instruments.

However, the world science timeline is more successful in communicating Manchester's role in scientific research and discovery. It really combines three timelines: general world events, significant scientific events worldwide, and Manchester scientific events. Significant Manchester creations are picked out and interpreted with objects and graphics. Among the highlights are drawing and surveying instruments, photographic shutters and dry plates, science teaching instruments and microscopy equipment.

The other part of this gallery is devoted to the lives of four famous Manchester scientists: John Dalton (1766–1844), James Prescott Joule (1818–1889), Ernest Rutherford (1871–1937) and Bernard Lovell (1913–). Each scientist has his own themed room and two 'Timeslices', the latter sections being the best attempt in the exhibition to contextualise Manchester science. For each of the scientific lives, there

A display in the Lovell room

are holographic reconstructions of two seminal events. Images of actors playing the scientists and their colleagues are imaginatively projected onto miniature room sets. We find out how Dalton gave scientific advice to Manchester's industrialists at the bowling green of the Dog and Partridge Inn in Stretford, and how the Manchester Literary and Philosophical Society supported Dalton's work. We discover how Manchester led the world in the practical application of science in the 1840s, and how its mastery of technology and instrument-making helped the young Joule to investigate the nature of heat and energy. We learn that Joule's work became vital to those who used and made steam engines, the power behind the industrial city.

An interactive in the Lovell room

We see how under Rutherford's dynamic leadership Manchester became an important centre for atomic research that launched the careers of many young physicists, and how he presented his revolutionary ideas and experiments on atoms to international conferences. Finally, we learn from Lovell himself how he moved from developing military radar during World War II to using radio waves for exploring deep space. From there he became the driving force behind the giant radio telescope at Jodrell Bank, which achieved immediate fame as the only installation that could track the first ever satellite (the USSR's Sputnik 1) in 1957. This is an excellent series of displays and perhaps it should have been more up-front rather than tucked away in the middle of the gallery.

The four themed rooms that detail the work of these scientists follow a formula. Each begins with a case of objects which contains apparatus or replicas used by the scientist in question. Panels of graphics and text then explicate the important theories and discoveries of the scientist, and a final section uses objects and graphics to show the importance of that work today. The centrepiece of each of the rooms is a hands-on interactive that reconstructs one or two of their experiments, and the back wall is devoted to a computer-generated animation that further illustrates their ideas and uses video to give some context. Still, the rooms are heavily focussed on fleshing out the content of the ideas and their relevance to present science in a manner that you would expect to find in a science centre. There is little insight into the scientists' personalities, and their social worlds are mostly glossed over. You do get a sense of how Dalton was motivated by religious faith and his memorial stone is used to good effect to show his importance in the local community. But these contextual moments are all too few for a gallery that purports to be about Manchester and its science. The model of scientific development used is a presentist one, measuring the value of the past scientific ideas in the light of their value today. Visitors get little sense of the scientific thinking at the time, or of social and political contexts, or the local factors that made Manchester a conducive environment for the gestation of these important ideas.

The gallery concludes by taking us up to the present in 'Science Today and Tomorrow'. An overhead conveyor belt displays objects that have been made possible by recent scientific research in Manchester. As the objects move round, a video for each of them shows Manchester scientists explaining their research and what motivates them. Here you get to see the latest robotic hand from the University of Salford and try it out in an interactive. You also get a good sense of the range of current research in Manchester and its practical uses: from helping to produce socks that never smell or harnessing the

The 'Science Today and Tomorrow' exhibit

power of the ocean's waves, to developing a drug that reproduces the scar-free healing experienced by the foetus in the womb. The area forms a very effective standalone exhibit.

In general, the gallery has rather too many objects which are not fully brought to life by the interpretation, and too many disparate sections with different briefs. It could have been more interactive, and it would have benefited from a clearer brief and fewer styles of interpretation to produce sections that were more coherent with one another. It has the hallmarks of a project in which design factors have taken precedence over research and interpretational concerns.

Many of the best objects and displays are lost in the mix and too much of the content is generic rather than focussed on Manchester's rich stories and characters. There are scripts available on seats in the gallery that contain the exhibition texts, with poor quality photocopies of letters, papers and manuscripts, but these are easily missed and they should have been interpreted in the gallery rather than appearing as what seems like an afterthought for the more curious visitor. The gallery does contain a fascinating array of objects and sources, and it succeeds in explaining complex topics in atomic theory, radio astronomy and energy conservation in a fun and engaging way, but it could have been so much more. It is not a world-class exhibition, and that is a shame in a city which could be one of the world's prime sites for showing the historical interconnections of scientific and industrial developments.

SOCIETIES
Manchester Astronomical Society

Kevin Kilburn

The Manchester Astronomical Society was originally formed in 1892 as the North Western Branch of the British Astronomical Association. The BAA was created as a national astronomical society in 1890 following the same pattern as the very successful Liverpool Astronomical Society that had been founded in 1881 but which, by 1890, was obliged to temporarily cease its activities until some years later owing to a financial crisis.

Initially peripatetic, meeting in various coffee houses in and around the Deansgate, St Ann's Square and King Street area, in 1903 the Manchester-based North Western Branch of the BAA became an independent society following the provision of a permanent meeting place at the Godlee observatory high above Whitworth Street, in the main building of the Manchester Technical School (now the Sackville Building of the University of Manchester).

A falling-out with the Principal at the Technical School led to the MAS removing in 1920 to 65 George Street and sharing accommodation in the former home of Manchester scientist, John Dalton, with the Manchester Literary and Philosophical Society. Shortly before World War II the MAS removed again to the newly built Central Library before returning to the Godlee Observatory in 1946. The Godlee observatory has been the home of MAS ever since.

The Godlee observatory was given to the city as a gift of Francis Godlee, a Manchester cotton millionaire and philanthropist. In 1900 he spent £10,000 to provide a two-storey wood-clad observatory tower cantilevered off the northern side of the Technical College. The lower rooms now provide a meeting room and annexed library and computer room for the MAS. The observatory itself is equipped with a rare twin-equatorial telescope, comprising an 8-inch refractor counterbalanced by a 12-inch reflecting telescope built by Grubb of Dublin. Sir Howard Grubb personally commissioned the instruments in 1902.

Only four twin-equatorials were ever constructed by Grubb; the first was ordered by the London spectroscopist William Huggins, the second by the then president of Liverpool Astronomical Society, Isaac Roberts, with which in 1889 he was the first to photograph the spiral structure of the Andromeda galaxy. At about the same time a third twin telescope was put on display at a Manchester exhibition but was dismantled and sent to observatories in Brazil and India. The Godlee

instruments were the fourth and last twin telescopes Grubb constructed to maximise the use of a small observatory. The refractor was designed for visual use, for lunar and planetary observation and the measurement of double stars. The reflector was intended for deep-sky photography. The two main instruments were complemented by a 6-inch diameter, f6, wide-field astrographic camera.

The Manchester Astronomical Society is one of the longest surviving Manchester scientific societies of the late-nineteenth century. It still holds weekly meetings at the Godlee observatory, every Thursday evening, and for more than a century has provided seasonal monthly public lectures, free of charge, commencing in September until March which are given by professional astronomers of the highest standing. New members and visitors are most welcome.

The MAS holds regular meetings on the third Thursday of the month from early autumn to late spring. They are held in the Renold Building and start promptly at 7.30 pm, and conclude no later than 9.30 pm. The Presidential Lecture is held in the Godlee Observatory, situated on the roof of Manchester University.

The address for the meetings is:
  Renold Building
  University of Manchester
  Sackville Street
  Manchester

Further details about MAS are published on http://www.mikeoates.org/mas/

Andromeda Galaxy, taken with a 300mm f5.6 lens and a Canon 350D DSLR. It is a stacked image of 5 x 60-second exposures combined

LONG REVIEWS

Derek Brumhead and Terry Wyke (eds.)
Moving Manchester: aspects of the history of transport in the city and region since 1700, Manchester: Lancashire and Cheshire Antiquarian Society, 2004, x+298pp, illustrations, maps, figures, tables, index. ISSN 0950–4699

Recent trends in transport history suggest that the field is at least catching up with – and might even be on the cusp of moving beyond – the sorts of theoretical concerns that have informed other kinds of historiography for some years. The sort of thing I have in mind is the greater emphasis being given to the way that people use, or 'consume', transport, with a consequential shift towards the analysis of cultural factors which inform, and in turn are informed by, changing patterns of mobility. This growing focus on how and why people have chosen to move themselves and their goods has also encouraged historians to think more about transport as a system of sometimes competing and sometimes complementary modes, and less about the individual types analysed in isolation one from another.

As they make clear in an insightful introduction, the editors of this valuable collection are not unaware of such trends, although almost without exception the subsequent 14 essays are more faithful to the older paradigm, favouring particular modes over system, questions of supply over use, and largely eschewing cultural factors in preference for those more traditionally associated with social and economic history, and industrial archaeology. No matter, for if the more recent historiographical fashions are to succeed in deepening our understanding of how mobilities are shaped by and sustain social power, we need to maintain, and indeed sharpen, analysis of the material constraints and opportunities represented by individual transport technologies. For the most part, these essays fulfil this role very well, and in some cases (notably those dealing with aviation, early road transport and river navigation) they cast a valuable light on subjects of which we are generally quite ignorant. There is also much to be said for studying the various elements of a region's transport system in some depth, particularly for that period before, say, the mid-nineteenth century, when it is arguable that traffic flows were far more constrained geographically than they were to become once the railways extended nationwide. Indeed, I am hard pressed to think of any other collection about the UK which addresses such a variety of transport modes in a regional context.

Waterways lie at the heart of three essays. Michael Nevell argues that archaeological evidence unearthed in Manchester over the last couple of decades points to a much greater importance for inland navigation in the growth of the city before the coming of the railways. David Vale takes this argument further by suggesting that the improved natural navigations, the Mersey and the Irwell, were themselves a good deal more important in relation to the Bridgewater Canal than most historians allow. Finally, Alan Findlow and Martin Whalley summarize the results of many years' documentary and archaeological research into the history of Bugsworth

Basin in north-west Derbyshire, from the late-eighteenth century a key trans-shipment point from plateway to canal for limestone and its derivatives, crucial materials for the building of Manchester and several of the region's industries.

Part of Vale's argument concerns the continued importance of road transport even as improvements were carried out to inland navigation, and Geoffrey Timmins builds on this point in a fascinating study of the evolution of Manchester's roads as they struggled to accommodate the increasing volumes of traffic generated by industrialization. Navigations and later railways may well have handled the greater volume of bulky goods carried over any distance, but they often relied on roads for local distribution as well as ceding to them a proportion of the inter-urban flows of lighter goods. Passenger traffic was important too, and Ted Gray's sketch of three generations of the Greenwood family underscores the role they played in providing increasingly affordable public transport in the nineteenth century, before municipalization and electric tramways entered the scene around 1900.

Railways inform a further three of the essays. Robina McNeil argues succinctly and persuasively that the design of the 1830 warehouse at Liverpool Street Station drew heavily upon earlier canal-side warehouses in Manchester and elsewhere. David Hodgkins draws upon his extensive and excellent biographical research to review the role of Edward Watkin, Manchester-born and -bred, in the world of mid- and late-Victorian railway management. And finally, Derek Brumhead develops a detailed study in the mould of J. R. Kellett's classic work, of the effects on the built and social environment of one of Watkin's major achievements, the construction of Manchester Central station.

In terms of the manufacture of transport equipment, the Manchester region is perhaps best known for its endeavours with regard to railways, but Nick Clayton and David George remind us respectively of a late-nineteenth industry (bicycles) and an early-twentieth century one (motor cars) which both enjoyed a modest success before the focus of production moved elsewhere, in both cases principally, of course, to the midlands.

More successful in the longer term has been Manchester's connection with aviation, not so much aircraft manufacturing (although there was some) as flight itself. R. A. Scholfield and Viv Caruana's essays offer two fascinating studies of the politics, national as well as regional, behind the location and operation of Manchester's airfields. Ringway – or Manchester International Airport as it is now more properly titled – is of course the culmination to date of the process of providing what seems to be the public's insatiable appetite for flying. But before the Second World War, and for that matter for sometime thereafter, aviation was a far from commercial proposition, requiring a good deal of faith and not inconsiderable subsidy to keep the planes in the air, not least by providing suitable facilities for take-off and landing. The generosity of earlier generations of Manchester's ratepayers in this regard contrasts strongly with the marked reluctance in the nineteenth century to subsidize the construction of railways.

Although Ringway now turns in a considerable financial profit for the benefit of its local-authority shareholders, it is a moot point whether either the regional or the national economy truly benefits once the horrendous environmental and other costs of aviation are taken into account. I wonder then whether in 50 years' time our descendants will be touring the Ringway Experience, admiring the skill and dedication of enthusiasts painstakingly restoring redundant charter planes parked alongside the largely silent Terminals One, Two and Three? John C. Fletcher and George Turnbull remind us that older forms of transport have met a similar fate. The former's account of canal restoration in the north-west chimes nicely with the interest by leading waterways historians such as Joseph Boughey and Mike Anson in the growth of waterways as a leisure industry, whilst the latter offers a useful account of the origins and collections of the Manchester Transport Museum.

In sum, this collection should prove a delight for any historian concerned with the social and economic history of Manchester and its hinterland, as well as for students of British transport more generally. Perhaps it will inspire some to undertake research into other aspects of the region's transport, such as municipal stage-carriage services, or the environmental effects of ever-increasing levels of mobility.

Colin Divall
Institute of Railway Studies and
Transport History, York

Ian Haynes
Hyde cotton mills (published by the author)
David Nelson
The Nelsons of Nelson: the story of a Lancashire textile firm and family 1881–1981 (published by the author)

The two towns of Nelson and Hyde are of roughly the same size and share a common involvement in the cotton industry. They were also towns where the leading cotton employers were associated with technological innovation. Two recent publications have added to our knowledge of the cotton industry in the two towns.

Ian Haynes has added a further instalment to his work on the Tameside cotton industry in *Hyde cotton mills*. This publication follows the format of his previous work in providing both a history and directory of the town's cotton industry and mills, as well as a series of short biographies of some of the key cotton entrepreneurs. The result is a welcome introduction to the cotton industry in Hyde with a particular emphasis on the Ashton family. Hyde played a leading part in the development of cotton during the first half of the nineteenth century with the Ashton family central to events. They were one of the most influential families in the Manchester region and were involved in the development of the Ship Canal and the establishment of the University. Liberals and Unitarians, their model village at Flowery Field attracted considerable attention and, according to Engels, this family was comparable with the Gregs of Styal.

Haynes offers an explanation for the growth of Hyde's cotton industry and in particular shows the significance of the combined spinning and weaving firm to the history of the town. Hyde was one of the first areas to develop power-loom weaving with handloom weaving virtually extinguished by the mid-1820s. However, with the growth of a specialist power-loom weaving sector after 1850 in North Lancashire, Hyde went through a very difficult period in the late-Victorian era. Cotton was rescued in Hyde by both Edwardian prosperity and the introduction of automatic looms and ring spinning. Ashton Bros' investment in new technology was to be important after 1920, giving the cotton industry in Hyde a competitive edge until the eventual decline and take-over by Courtaulds in the 1960s.

David Nelson's history of the Nelson family is part personal memoir and part history. Sir Amos Nelson, his father, was one of the key figures in the history of the cotton industry in the first half of the twentieth century. The growth of the family firm in late-Victorian Nelson is already known. Sir Amos followed his father, a former weaver and overlooker, in creating one of the most dynamic firms in North Lancashire. The account offers a particularly interesting insight into the family dynamics in the firm, describing the different economic roles of Sir Amos, his father and his sons. Two chapters are devoted to the 'more looms' crisis and Sir Amos's role in the Nelson lockout of 1928. A further chapter covers the wider social life of the family including Sir Amos's decision to commission Sir Edward Lutyens as architect for his new home, Gledstone Hall. The firm and the family, which was so influential in the political, social and industrial life of the town of Nelson deserves far greater attention from economic historians. The firm was always in the vanguard of technical change, investing in semi-automatic looms at the height of the 1930s depression, whilst also developing artificial silk. It remained successful into the 1960s when again, like Ashton Bros, it was taken over by Courtaulds.

These two publications bring together accounts of two families in the history of the Lancashire cotton industry, both relatively neglected and in need of further exploration. Ian Haynes and David Nelson have raised the profile of their subjects and it is to be hoped that others will be encouraged to build on their work and explore the history of cotton in Hyde and Nelson.

Alan Fowler

Michael Nevell and John Walker
The archaeology of twentieth-century Tameside, Manchester: Tameside MBC/Manchester University Archaeology Unit, 2004, 123pp. ISBN 1–871–32429–7

This book by Nevell and Walker of the Manchester University Archaeology Unit covers what they call the archaeology of Tameside from 1870 to 2000; it is, however, much more than that. It is a profile of the area and is quite comprehensive, even more so when it is seen as the last of a series of seven on the borough. The profile covers the growth of the area from 1870 and the changes that have occurred to both the industrial and rural landscapes. There is a chapter on the archaeology of local government which traces reforms and the buildings erected as a result. There is also a section that deals with the question of the definition of archaeology especially when it is used in connection with the post-industrial revolution period. One important point to recognise is the support given by Tameside MBC in the production of the book and the survey work that lies behind it. It is a pity that more local authorities are not quite as generous.

This is a serious contribution to the history of the area, which, like so much of our region, lacks good analysis of the events that have shaped it. There is a good scattering of maps and graphs to illustrate the changes the authors describe. One aspect of the layout of the book I found especially pleasing is that throughout they use cartouches to describe events and people tangential to the line of the text. One of these is a thumbnail sketch of Winifred Bowman, the Ashton local historian who died in 1984. It is a nice touch. The cartouche is such a good idea that I might borrow it in the future.

Nevell and Walker are at their best when they are dealing with the meat of their subject, archaeology. They are on less firm ground when talking of more peripheral subjects such as the development of local government. Their sub-title for the work 'From lordship to local authority: the archaeology of the later industrial period, 1870–2000', and some of the text, suggest a less than radical change in local government by the middle of the nineteenth century. The fact that the Ashton Vestry was still meeting in 1911 is of little consequence, especially as the authors point out the town had a self-governing local board by 1827 and was a full municipal borough by 1847. To see any parallels with medieval forms is to ignore these radical changes wrought in local government. Vestiges of medieval institutions such as 'lords of the manor', 'lord lieutenants' and 'justices of the peace' are, as Walter Bagehot stated over 100 years ago, 'new wine in old bottles'.

On the whole, though, it is a fluent book. That said, Chapter One is a little bit difficult to digest with its description of the current thoughts on the subject of archaeology and its limits. I prefer 'the bear of small brain' approach to labelling things although I recognise someone has to do it. One of the consequences of the very academic opening chapter is that one feels that Nevell and Walker are 'shoe horning' some of their profiles into what they are keen for us to accept is archaeology. I think there are a few geographers, urban historians and the like who might feel a little redundant. The problem is not what they put in their book, but that the use of the term 'archaeology' jars a little in places – this comes out most obviously in the chapter on local government. And while I am having a little niggle, on page 58 they state that the social status of the housing of Hyde increased to the south and south-west. They are surely stating the obvious – the status of the housing of every town in England tends to increase to the south and south-west because of the way the wind blows most of the time – why are Openshaw, Gorton and Miles Platting north and north-east of Manchester? They also use the term 'the archaeology of the industrial revolution' on page 3 when I think they mean the archaeology of the industrial period. Revolution by its very definition must cover a limited period of events or it is looking a bit like evolution.

In spite of these reservations I think we have to thank the Archaeology Unit and the backing of Tameside MBC for this and the first six volumes on the history, development and archaeology of the area. It is a worthy contribution and should be, along with its six companions, on the bookshelf of any self-respecting local historian.

Steve Little

Alan Fowler

Lancashire cotton operatives and work, 1900–1950: a social history of Lancashire cotton operatives in the twentieth century, Aldershot: Ashgate Publishing, 236 pp., illustrations and tables. ISBN 0–7546–0116–1

Surely not another book on the history of the Lancashire cotton industry! This might be the reaction of many readers, when faced by this volume and numerous similar books that load library shelves. Within the last decade or two, there have been several major studies – John Singleton on Lancashire's decline, Geoff Timmins on Lancashire as a region, an edited collection by Mary Rose, Roger Lloyd-Jones's and Mervyn Lewis's analysis of the industry structure, Douglas Farnie on Lancashire's role in world trade – to say nothing of the journal literature. Yet it is a sign of the significance and scope of the present volume that it has little difficulty in claiming our attention. This is because economic and business history – certainly as regards Lancashire – has always been a conservative profession, much vexed by the issues of production, profits, entrepreneurship, technology, and business structure, yet devoting relatively little attention to social aspects. With the exception perhaps of J. K. Walton and H. A. Turner, the only historians to have devoted systematic and sustained attention to labour and social issues in the cotton mills have been Alan Fowler (and his colleague at Manchester Metropolitan University, Terry Wyke).

This book contains much that is familiar. It sets out the (by now) well known story of the rise and decline of a major British industry in roughly chronological fashion, beginning with an introduction on the rise and fall of King Cotton. But this is interwoven with a novel treatment of key labour themes. The first of these is the role of the trade unions, which is the subject of the first main chapter. Fowler shows how the labour landscape was dominated before 1914 by three of the largest amalgamations of the British labour movement – the Weavers' Association, the Society of Card and Blowing Room Operatives, and the Association of Operative Cotton Spinners. The first was the second largest union in the country; the last was the richest, peopled by the 'barefoot aristocrats' (those relatively privileged artisans, who trod the mill floors without clogs). The unions were sectional, split between weaving and spinning, with the former trade conducted in north Lancashire and the latter in the south. The unions were also split by gender: most spinners were men and, although this situation was the opposite in weaving, nevertheless most weaving trade union officials were male. This complex structure is described lucidly by Fowler, who shows that the fragmentation of the union movement (and the growth of powerful employers' associations) did not prevent the unions from forging ahead before 1914. This was achieved by collective bargaining, which resulted in mutually agreed wage lists. Nor did sectionalism prevent the unions from launching more broadly-based movements, such as the United Textile Workers' Association, which was intended to further factory reform.

The exclusion of women from union power was ironic, given that, unlike other industries (such as steel and engineering), most cotton workers were female. This gave the Lancashire labour force its special character, at least until 1914. In Chapter

Three, Fowler paints a fascinating portrait of a highly literate labour force, which had its own newspaper (the *Cotton Factory Times*), and was relatively prosperous. This was achieved by the kind of hard work that is now a distant memory: up at 5 am for a working day that did not end until 5.30 pm. Child labour was a crucial input into higher living standards, as was the contribution of female workers, who still found enough energy to raise families and keep the house clean. The development of holidays in Blackpool and leisure pursuits such as football, cricket, and the music hall were the other essential ingredients in this picture. The inter-war period was tougher, with cotton trade union membership falling from about 400,000 in 1921 to half that by 1939. In the face of wage cuts, strikes and 40 per cent of the workforce on the dole, the spinners' privileged position weakened and the weavers became embroiled in productivity disputes. Even here, though, it was not all doom and gloom, with the rise of international trade unionism and the launch in Manchester, in 1924, of an international federation for textile workers' associations.

Why the cotton trade unions were not more effective politically in the inter-war depression is explained in a chapter on trade unions and politics. It was not through lack of application or hard work, since many renowned trade unionists came from poor backgrounds and achieved success only through a part-time technical education. This could only be experienced after working all day in the mills (although one resourceful weaver, Tom Shaw, is reported to have learned foreign languages by pinning lists of verbs on his loom). Moreover, union officials were only appointed after rigorous examinations that could take days. However, the main object of most of the tests was to show whether candidates had the mathematical ability to navigate through the complexities and mysteries of the wage lists. According to Fowler, the result was that the cotton unions produced excellent technicians, but men lacking political acumen. As the Labour party blossomed in the early-twentieth century, the cotton unions had virtually the same number of MPs in 1945 (three!) as they had in 1906.

A major contribution to the literature of the cotton industry is the chapter on health and safety – until recently a subject ill-served by historians. Cotton was known as a relatively safe industry, but nevertheless it had its dangers. For example, card room workers could suffer and die from byssinosis (cotton dust respiratory disease); while spinners risked a painful and lingering death from mule spinners' cancer (the euphemism for scrotal cancer caused by engineering oils). More general risks were from accidents, shuttle-kissing, and the effects of steaming (raising the heat and humidity) in the weaving sheds. Each risk is expertly described, with the emphasis on how the trade unions doggedly tried to improve matters, even though – as might be expected – their main concern was wages and working hours. The long-term record of the unions is shown to have been mixed, but Fowler is surely right to stress that the reason for this was not the failings of the unions, but the major obstacles they faced (such as employers' intransigence) in a declining industry.

Like the industry it describes, this study has some idiosyncrasies. The choice of the dates 1900 and 1950 is not explained and the 1950 cut-off means that some subsequent aspects are simply not treated. For example, many interesting developments in the health and safety field occurred after the 1940s. On the other hand, much information is included on the late-nineteenth century. I also pondered why fining (as a punishment for poor quality work) was included in the health and safety chapter. Bracketed references are used, which means that apart from the annual reports of the unions, no cotton manuscript collections or key documents are cited, either in Manchester or further afield at places such as The National Archives. In the absence of a 'note on sources', Wyke's and Rudyard's *Cotton: a select bibliography* (1997) is an unfortunate omission. Although the international labour movement is given some attention, the focus is resolutely on Lancashire and the wider significance of many of the social aspects illuminated here are not drawn out as fully as they might be.

Nevertheless, this book is clearly set out, easy to read, has an excellent bibliography, and its conciseness will have attractions for many readers. The subject of the trade unions in the cotton industry has traditionally not been an easy subject for historians or the general reader to penetrate, but their task will be a lot easier now. There may be many other books on Lancashire cotton, but this will hold its place as the standard work in its field. It is an important contribution to the literature of the Lancashire cotton industry that has been needed for some time.

Geoffrey Tweedale
Manchester Metropolitan University